T0122198

Studies in Big Data

Volume 110

Series Editor

Janusz Kacprzyk, Polish Academy of Sciences, Warsaw, Poland

The series "Studies in Big Data" (SBD) publishes new developments and advances in the various areas of Big Data-quickly and with a high quality. The intent is to cover the theory, research, development, and applications of Big Data, as embedded in the fields of engineering, computer science, physics, economics and life sciences. The books of the series refer to the analysis and understanding of large, complex, and/or distributed data sets generated from recent digital sources coming from sensors or other physical instruments as well as simulations, crowd sourcing, social networks or other internet transactions, such as emails or video click streams and other. The series contains monographs, lecture notes and edited volumes in Big Data spanning the areas of computational intelligence including neural networks, evolutionary computation, soft computing, fuzzy systems, as well as artificial intelligence, data mining, modern statistics and Operations research, as well as self-organizing systems. Of particular value to both the contributors and the readership are the short publication timeframe and the world-wide distribution, which enable both wide and rapid dissemination of research output.

The books of this series are reviewed in a single blind peer review process.

Indexed by SCOPUS, EI Compendex, SCIMAGO and zbMATH.

All books published in the series are submitted for consideration in Web of Science.

Victoria N. Ostrovskaya · Aleksei V. Bogoviz
Editors

Big Data in the GovTech System

Editors
Victoria N. Ostrovskaya
Center of Marketing Initiatives
Stavropol, Russia

Aleksei V. Bogoviz
Independent researcher
Moscow, Russia

ISSN 2197-6503 ISSN 2197-6511 (electronic)
Studies in Big Data
ISBN 978-3-031-04905-7 ISBN 978-3-031-04903-3 (eBook)
https://doi.org/10.1007/978-3-031-04903-3

This Springer imprint is published by the registered company Springer Nature Switzerland AG
The registered company address is: Gewerbestrasse 11, 6330 Cham, Switzerland

Big Data In The GovTech System: Scientific Vision and Modern Empirical Experience (Introduction)

Big Data belongs to Industry 4.0. It emerged and became widespread in various areas of business around the world under the influence of the Fourth Industrial Revolution. As a scientific category, Big Data is a relatively new phenomenon in science, which is in the process of discussion. There are interpretations of it as:

- Management object: Big Data as an information base (large data array) for making managerial decisions (Rehman and Noman 2021; Sandhu 2022; Zhao and Zhou 2022);
- Management process: Big Data as a data collection process and an analytical procedure for transforming "raw" statistics into a knowledge base through establishing cause-and-effect relationships between data (Gökalp et al. 2022; Vranopoulos et al. 2022);
- Management technology: Big Data as a technology for intelligent management decision support based on deep automated analysis of large-scale information (Glavind et al. 2022; Guo et al. 2022).

In practice, Big Data is actively used and is getting more and more applications, which are constrained and associated with reduced efficiency due to insufficient scientific and methodological support. Therefore, the scientific study of the theory of Big Data and its harmonization with existing applied solutions is of high relevance.

In this book, we seek to form a systematic understanding of Big Data and, therefore, are guided by their broad interpretation in the materials of the OECD (2022), which, like other international organizations, plays an important role in building a bridge between the theory and practice of Big Data. Big Data, in this book, is understood as the whole set of information (large volume), the process of its transformation into digital data, automated processing and analytics, as well as the use of intelligent management decision support.

GovTech is another key category of this book, the essence of which is the high technology of the state apparatus in the economy and its activity. Its conceptual foundations are also in the process of formation and discussion, and in general, there is a lack of scientific research on the topic of using Big Data in the GovTech. Therefore, in this book, we refer to the World Bank (2022) definition, which encompasses three manifestations of the GovTech: digitally delivered and therefore open, convenient, and accessible government services; transparent and automated state regulation of the economy; effective digital government system.

The functional load of the GovTech requires it to be sound and highly effective, and therefore it must be based on reliable data, as noted in GovTechLab (2022). This book is aimed at studying the empirical experience of using Big Data in the GovTech system, as well as its scientific rethinking and the development of promising applied solutions for its improvement. Three manifestations of Big Data in the GovTech system are revealed in three parts of this book, the originality of which lies in the systematic study of Big Data in the GovTech, which ensures the contribution of the book to the literature, as well as in a detailed study of modern experience and cases of using Big Data in the GovTech system on the example of Russia, which determines the practical significance of the book. The experience of Russia is largely universal, which makes the author's conclusions and recommendations also applicable in other countries, especially in developing countries (for example, in the BRICS).

Part I deals with the GovTech in the provision of high-tech educational services based on Big Data. State universities dominate in Russia and, in general, the functioning and development of the higher education market are regulated by the state. To develop human potential and support the "knowledge economy" in Russia, higher education services are provided on a massive scale based on funding from the state (mainly from the federal) budget. Therefore, they conditionally represent a special type of state (public) services, in the provision of which Big Data is actively used.

Part II is devoted to studying the experience of state regulation of the economy by industry using Big Data in the GovTech. The experience of service organizations and the service sector in general, with special attention to financial services, as well as the experience of the oil and gas industry are taken into account. The third part identifies the digital divide and defines the prospects for overcoming it with the help of the GovTech based on Big Data. At the same time, various manifestations of the digital divide are taken into account: in the level of training of digital personnel in the labor market, in the availability of public digital infrastructure for various social categories, in the digital development of territories (regions) within the country.

This book is intended primarily for academics studying the GovTech. In the book, they will find a holistic scientific vision of the GovTech, as well as a systematized modern empirical experience of using Big Data in it. The secondary target audience for this book is the GovTech system. For them, the book offers applied developments and the author's recommendations of a scientific and methodological nature to improve the practice of using Big Data in the GovTech system. Due to the systematic nature of the book, it is a practical guide to improving the GovTech system based on the expansion and efficiency of using Big Data.

References

1. Glavind, S. T., Sepulveda, J. G., Faber, M. H. (2022). On a simple scheme for systems modeling and identification using big data techniques. *Reliability Engineering and System Safety*, 220, 108219. https://doi.org/10.1016/j.ress.2021.108219.
2. Gökalp, M. O., Gökalp, E., Kayabay, K., (...), Koçyiğit, A., Eren, P. E. (2022). A process assessment model for big data analytics. *Computer Standards and Interfaces*, 80, 103585. https://doi.org/10.1016/j.csi.2021.103585.
3. GovTechLab (2022). Big Data. URL: http://govtechlab.org/big-data/ (Accessed: 30.01.2022).
4. Guo, Y., Chen, J., Liu, Z. (2022). Government responsiveness and public acceptance of big-data technology in urban governance: Evidence from China during the COVID-19 pandemic. Cities, 122, 103536. https://doi.org/10.1016/j.cities.2021.103536.
5. OECD (2022). Big data: Bringing competition policy to the digital era. URL: https://www.oecd.org/competition/big-data-bringing-competition-policy-to-the-digital-era.htm (Accessed: 30.01.2022).
6. Rehman, F. U., Noman, A. A. (2021). China's outward foreign direct investment and bilateral export sophistication: a cross-countries panel data analysis. *China Finance Review International*. https://doi.org/10.1108/CFRI-04-2020-0040.
7. Sandhu, A. K. (2022). Big data with cloud computing: discussions and challenges. *Big Data Mining and Analytics*, *5*(1). https://doi.org/10.26599/BDMA.2021.9020016.
8. Vranopoulos, G., Clarke, N., Atkinson, S. (2022). Addressing big data variety using an automated approach for data characterization. *Journal of Big Data*, *9*(1), 8. https://doi.org/10.1186/s40537-021-00554-3.
9. World Bank (2022). What is GovTech. URL: https://www.worldbank.org/en/programs/govtech (Accessed: 30.01.2022).
10. Zhao, Y., Zhou, Y. (2022). Measurement method and application of a deep learning digital economy scale based on a big data cloud platform. *Journal of Organizational and End User Computing*, *34*(3). https://doi.org/10.4018/JOEUC.20220501.oa1.

Contents

Digital Divide and Its Bridging with the Help of GovTech Based on Big Data

GovTech in the Provision of High-Tech Educational Services Based on Big Data

On the Need and Opportunities for Digitalization of the Educational and Methodological Support of the Educational Process in the Context of Improving Its Quality Indicators

Yuliya Ye. Golovina and Yury Yu. Grankin

Abstract Purpose: to consider the possibility and expected effectiveness of introducing digitalization tools into the educational process using RPA technology to improve the quality indicators of educational and related processes. Methodology: to achieve this goal, we use the capabilities of system analysis, which allows us to combine and investigate the factors of the external and internal environment of the university that determine the quality indicators of the educational process, as well as the need and prospects for the introduction of digitalization. In turn, the synthesis of the data obtained and the conclusions made it possible to identify the main tool of digitalization and justify our choice. Moreover, the author's argumentation was based on the results of qualitative research methods: focus groups and expert assessments, as well as methods of systematization and visualization of the data obtained. Findings: the analysis of the requirements of legislation in the field of education, the realities and trends in the development of information technology, as well as the actual practice of organization of business processes in an educational organization, allowed us to conclude that the introduction of digitalization in the educational process implies, on the one hand, a positive effect. Thus, scientific and pedagogical workers will no longer have to devote a significant part of their working and free time to the preparation of teaching materials, which will lead to an increase in the creative potential of teaching staff, motivation to work, the level of dissatisfaction, stress from paperwork and intra-university bureaucracy will decrease. The coordinating and controlling functions of the university management, readiness for all kinds of control and supervisory measures will also be optimized. On the other hand, the study revealed the possible risk levels of implementing RPA and the strength of their negative impact on business processes. Originality: the presented study is characterized by a high degree of originality, which is due to the nature of the study itself: the authors set a goal, the achievement of which involves solving specific pressing problems, rather than theoretical analysis and obtaining hypothetical conclusions.

Y. Ye. Golovina (✉) · Y. Yu. Grankin
Pyatigorsk State University, Pyatigorsk, Russia
e-mail: julia.golovina@pgu.ru

Y. Yu. Grankin
e-mail: grankinj@pgu.ru

V. N. Ostrovskaya and A. V. Bogoviz (eds.), *Big Data in the GovTech System*, Studies in Big Data 110, https://doi.org/10.1007/978-3-031-04903-3_1

The research is based on specific processes, their subjects and objects in the organization under study, applied research methods that are situationally suitable for this particular problem. Conclusion: the presented analysis made it possible to argue for the need to introduce digital transformation technologies into the work of modern organizations, including the field of education since this will largely optimize routine processes at the university, improve the quality and efficiency of the formation of methodological and organizational components of the educational process.

Keywords Informatization of education · Science and education of the Russian federation · Educational and methodological support · Digitalization of education

JEL Codes O3 · O310 · O320 · O350

1 Introduction

In the era of total digitalization, there are still separate industries whose "totality" of digitalization is difficult, since these processes have often not yet affected not only the substantive side of the functioning of industries but even the organizational and supporting factors of their existence [1], p. 770.

One of such branches in the Russian economy is education, in particular higher education, where, despite many "digitalized" publications, conferences, courses, elementary issues of organization and provision of the educational process have not yet been resolved [2]. This problem is especially acute in regions that neither materially nor meaningfully "do not keep up" with the capital's universities and trends in the field of digitalization in general [3, 4], p. 753.

The Federal State Budgetary Educational institution of Higher Education "Pyatigorsk State University" for almost 80-year history has gone from a secondary educational institution of pedagogical orientation to one of the leading linguistic universities of the country (PGLU), and then the classical university (since 2015—PSU), which currently implements more than 120 basic educational programs in more than 70 specialities and areas of training of secondary vocational and all levels of higher education. In addition, today the University is a centre of attraction and development of additional education, scientific research, practice-oriented project activities, various student movements, etc., and is also a platform for social and political events at levels from local to international. It can be said that today the Pyatigorsk State University is an example of a good regional university that provides the satisfaction of the needs for basic and additional professional equipment at all its levels for residents of the home region, subjects of the North Caucasus Federal District, other regions of the Russian Federation and countries of the near and far abroad.

Despite the relatively small size of the university throughout Russia, the university has developed almost all business processes characteristic of an educational organization, but not all of them are fully covered by the processes of digitalization.

2 Materials and Method

To achieve this goal, we used the capabilities of system analysis, which allowed us to combine and investigate the factors of the external and internal environment of the university that determine the quality indicators of the educational process, as well as the need and prospects for the introduction of digitalization. In turn, the synthesis of the data obtained and the conclusions made it possible to identify the main tool of digitalization and justify our choice. Moreover, the author's argumentation was based on the results of qualitative research methods: focus groups and method of expert assessments, as well as methods of systematization and visualization of the data obtained.

3 Results

The full-fledged effective functioning of the university is ensured by the balanced implementation of business processes in the educational organization [5]. The specificity of an educational organization is that, with all the importance of such business processes as "finance", "marketing", "economic support", "personnel management", etc., in our opinion, the fundamental ones are still business processes of an educational and scientific nature and specifically the business process "Educational, organizational and methodological support of educational activities", which is the key in the block of educational processes (Fig. 1).

In the analyzed university, this business process is defined and effective from the point of view of its content (compliance with the requirements of the legislation of the Russian Federation at all levels, local regulations of the university, relevant intellectual parameters of the implementation of educational programs), however, it is rather poorly digitized, which often leads to excessive spending of time, labour and intellectual resources when updating, correcting and controlling the form and content of the business process "Educational, organizational and methodological support of educational activities".

At the moment, the IT infrastructure is used point-by-point (for example, in the process of planning the educational process—"HPE Plans Software Package + SPO" MMIS laboratories; in the process of providing educational literature—the MEGA Pro platform, etc.), without linking together the whole complex of elements of the business process "Educational, organizational and methodological support of educational activities". The analysis of the existing state of this business process at the university, as well as existing IT technologies, allows us to conclude the possibility and expediency of using RPA (Robotic Process Automation) technology, i.e. such an approach to business process automation in which the software product simulates human actions with information systems [7]. Based on the essence of this technology, we can say that its implementation should lead to:

- reduction of costs for performing routine operations;

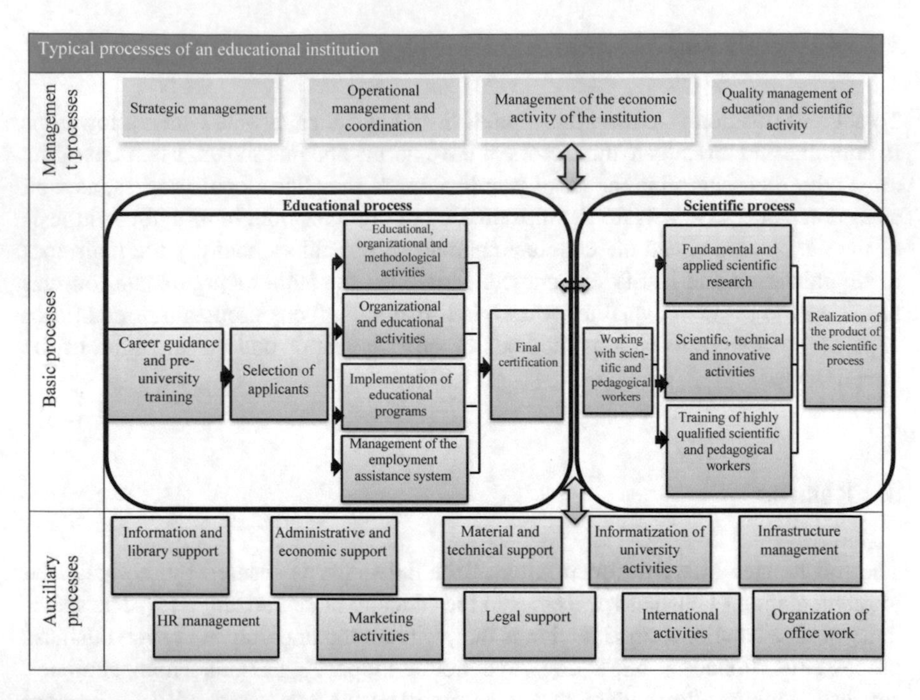

Fig. 1 Business processes of an educational organization. *Source* Compiled and developed by the authors based on the data of [6]

- reduction of the number of errors in processes due to the reduction of the influence of the human factor, improving the quality and speed of their implementation;
- reduction of various kinds of risks for the organization;
- shifting the focus of employees to performing intellectual and creative tasks [8].

In our case, the choice of this technology is due to the content of the business process "Educational-organizational and educational-methodological support of educational activities": most of the work takes the formation of curricula and educational-methodological support, consisting of work programs of disciplines, methodological materials, evaluation funds, in the process of which an array of specific data contained in federal state educational standards, curricula, information and library systems, etc. is used. In fact, the preparation of these teaching materials is a routine process, from most of which it is possible to painlessly exclude a person just with the help of RPA technology, leaving for scientific and pedagogical workers only the substantial part of the materials as a creative process of the teacher [9].

Teaching materials have only 1 creative section—"The content of training sessions (lectures, seminars), which depends on pedagogical skills, intellectual component and author's methodology. Most of the methodological materials can be formatted as a constructor just with the help of RPA technology from prepared reference books with the following data: "Forms of current control", "Forms of intermediate control", "Evaluation criteria", "Material and technical support", etc. These reference books should be formed by the relevant departments, as well as by the departments responsible for these data".

Such sections as the volume of discipline in hours/credits, semesters of implementation, forms of control, formed competencies are already contained in working curricula, and therefore, within the framework of the application of RPA technology, they can be automatically included in teaching materials, taking into account the criteria selected for specific materials.

The same applies to the section "Information and library support": with the help of the technology being implemented, the necessary information and library resources will be automatically selected and added to the educational and methodological materials.

Thus, the teaching materials will be the result of assembling, one might say, a constructor, in which, as a result, the teacher will need to add only his own author's component in the form of the content of training sessions. In addition, an important argument should be that digitizing the process of forming educational and methodological materials in this way will also lead to uniformity of all the materials received, which is not always possible in standard text editors.

From the point of view of the organizational component of this business process, it can be noted that the chosen technology will make it possible to efficiently, systematically and flexibly form the schedule of training sessions, intermediate and final attestations: with the help of pre-formed reference books, working curricula, the schedule will automatically receive information about the equipment and capacity of classrooms, the number of classes, as well as the number of study groups, which will simplify the organization of the educational process, especially when it comes to inter-faculty classes, in-line lectures, etc. [10].

It is obvious that the introduction of digitalization technology in a classical and rather an archaic university will be perceived ambiguously [11, 12]. Thus, one of the most important barriers may be the human factor, namely, resistance to changes on the part of both scientific and pedagogical staff of the university and administrative staff responsible for educational and methodological support, IT infrastructure of the university, as well as economic support, as it will require the formation and constant monitoring and correction of"reference books" used by the RPA product for the formation of educational and methodological materials.

In addition, it will be necessary to create so-called "crutches" to "unite" all the technologies already used at the university, which can also cause dissatisfaction and resistance of those responsible.

Thus, barriers may arise both on the part of users (scientific and pedagogical workers, especially older ones) and performers (sabotage and/or lack of competence of employees of the Department of Informatization, Educational and Methodological Department).

In addition, the material and technical base can also become a barrier—the existing material and technical support are not always able to ensure the correct operation of new software, in particular RPA technology.

This barrier will be associated with another type of problem—financial. Updating of material and technical support, development and implementation of new technology will require financial resources, which always causes certain difficulties in a state educational institution.

The most unpredictable barrier may be the so-called evolution of federal state educational standards and the regulatory framework on the implementation of higher education and secondary special education programs—these documents regulate, among other things, the educational and organizational, and educational and methodological aspects of the educational process, and therefore changing them will require constant adjustments in the RPA configured for the university.

Despite many barriers, the strength of potential drivers is quite significant. So, again, people are an important driver, only now the human factor can be considered from the perspective of proactive employees and structural units. According to one approach, such people can be viewed from three positions:

(1) pioneer people (initiators and technical executors of the technological initiative: in our case, the head of the informatization department, who is really interested in introducing elements of the digital transformation of the educational process);
(2) people are champions (representatives of the leaders who support the pioneers both morally and financially: in our case, the vice-rector for academic policy, quality control of education and informatization, since digitalization processes using RPA technology can increase efficiency and largely optimize the functional areas under his responsibility);
(3) Team X (people and departments who are ready to test the technology and make constructive adjustments through highly effective feedback: in our case, the Department of English and Professional Communication, as well as the Department of Creative and Innovative Management and Law).

An equally important driver is the permanent readiness of a higher educational institution for any kind of external and internal inspections, for example, state, public, professional and public types of accreditation, activities of control and supervisory authorities (Rosobrnadzor), in which educational and organizational and educational and methodological support is checked both online on the university's website and during on-site field inspections. In addition, there is the so-called "Spider" system and its subsequent versions, which, for control and supervisory authorities, constantly monitors the official websites of universities, and if there is a discrepancy, including in the aspect of educational and methodological support, it forms the basis for bringing the university to administrative responsibility.

Taking into consideration the fact that the business process "Educational, organizational and methodological support of educational activities" is the basic one in the educational process of the university, and the educational process, in turn, is the key one in the functioning of the entire educational organization, when analyzing possible risks, the entire possible range of areas of influence of the proposed technology on the business activity of the university was considered.

To determine the possible risks of implementing RPA and the strength of their impact on business processes, a working group was created consisting of employees of the Department for the Formation and Evaluation of the Quality of Education (4 people, including the author of the work), as well as the previously mentioned people who are part of the drivers of successful technology implementation (pioneers, champions, team X—only 6 people). The members of this working group, indeed, constantly interact in the process of introducing technical and educational innovations at the university.

Based on the results of brainstorming and analysis of the responses received, the following possible risks from the introduction of RPA technology were formulated (on the left there is a graphic image of the risk in the form of a circle of a certain colour, used further in the risk matrix to visualize the risk assessment):

- ◍– rejection/dissatisfaction of the team, which may eventually lead to sabotage of the introduction of RPA technologies;
- ◍– time risk—the probability of not meeting the deadlines by the time the materials need to be updated at the university and on the university's website following the requirements of the legislation;
- ●– overspending of resources—the initial overestimation of oneself and underestimation and poor forecasting of financial, time and intellectual costs;
- ○– to develop the wrong product that is initially required to solve the task (due to the appearance of problems in mutual understanding in the customer-contractor relationship, that is, difficulties in translating "from Russian into informational" language and vice versa);
- ◍– inability to correctly introduce the product into the daily activities of scientific and pedagogical workers from the position of the users themselves (lack of competence to work with software).

Thus, based on the vision of 10 interested employees of the university who were included in the working group, the risks of implementing RPA technology for the business activity of the university were assessed by two parameters: the probability of risk and the level of potential consequences. The final assessment of each of the identified risks is made up of the product of the values of probability and consequences. In Fig. 2, you can see a matrix of potential risks with the designation of the strength of the risks and the positions of the risks identified by us with the help of circles of various colours.

So, risks 1–3 have fallen into a high-risk zone. This fact indicates the need for special control by senior management in these areas of risk occurrence since the manifestation of these risks can lead to a violation of labour discipline, waste of resources, as well as the emergence of administrative responsibility due to violations of the requirements of federal legislation in the field of posting information on the official website of the university.

Risk No. 5 has fallen into the medium risk zone, which indicates the need for additional training and consulting of scientific and pedagogical workers on the use of new technology, which will help prevent the occurrence of the risk.

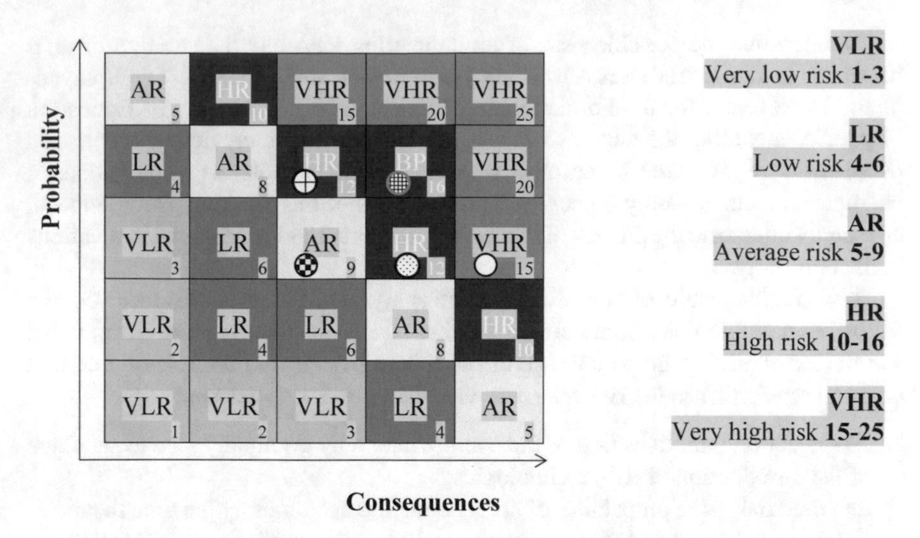

Fig. 2 Potential risk assessment matrix. *Source* Developed and compiled by the authors

The most serious, and in our classification, a very high risk, was risk No. 4. This fact indicates the need for more thorough and detailed attention to the implementation of RPA technology at the stages of conception (planning) and organization. Only a balanced approach to creating an IT product using RPA technology will allow the process of digital transformation of the business process to be carried out constructively. In our case, if this risk occurs and manifests itself, the entire process of implementing RPA technology will be a failure and irretrievably costly.

The successful introduction of RPA technology into the business process "Educational, organizational and methodological support of educational activities" at PSU, taking into account all barriers, proactive study of risks and the use of the power of drivers, will have a multifaceted effect from the point of view of participants in this process (subjects) [13], p. 24.

Speaking about the main participant-the user of the technology—scientific and pedagogical workers—it should be noted an unambiguously positive expected effect, which will manifest itself if all difficulties and risks are levelled. Thus, scientific and pedagogical workers will no longer have to devote a significant part of their working and free (more importantly) time to the preparation of teaching materials. Effective preparatory work (with proper definition, planning and execution) will allow you to create the necessary reference books at the initial stage and free teachers from routine work, which, as observations show, leads to emotional and professional burnout.

As a result, the creative potential of teaching staff increases, motivation to work, the level of dissatisfaction, stress from paperwork and intra-university bureaucracy decreases.

Another group of participants who should feel the positive effect of the introduction of RPA technology is the university administration. This form of transformation makes it possible to make the business process under study transparent: any responsible employee at any time will be able to see the process, statistics of work in the system, monitor the holistic picture of the implementation of the business process "Educational, organizational and methodological support of educational activities" at the university, and not collect data using administrative acts. Thus, the coordinating and controlling function of university management will be optimized. Another positive manifestation is the readiness for all kinds of control and supervisory measures, which were discussed above: the university management is always interested in full compliance with all requirements in order to avoid measures of administrative responsibility.

Of course, the university management will see the effect from the position of scientific and pedagogical workers: pedagogical workers as performers of a key function in an educational institution should feel an increase in the quality of life (professional and in general), which, of course, has a positive effect for the university as a whole.

4 Conclusion

The analysis carried out during the preparation and writing of this work once again confirmed the need to introduce digital transformation technologies into the work of modern organizations, including in the field of education. Digitalization should cover all business processes in an educational institution, but it is extremely important to introduce effective technology into the fundamental business process at the university "Educational, organizational and methodological support of educational activities", since this will largely optimize routine processes at the university, improve the quality and efficiency of the formation of methodological and organizational components of the educational process [14], p. 65. The RPA technology we have chosen, despite the ambiguity of the attitude towards it in the professional environment, is the optimal solution in our case, since it allows us to strengthen the creative component of teachers' work, increase their motivation, as well as streamline and improve the quality of teaching materials at the university, allowing to fulfil the requirements of the legislation in the field of education as much as possible.

References

1. Belskikh, I. E., Boriskina, T. B., Samsonova, E. V., Mershiev, G. A., & Tsygankova, V. N. (2018). Marketing of Russian universities in the education market: In search of growth alternatives. *Economics and Entrepreneurship, 7*(96), 770–772.
2. The main trend of Russian education is digitalization. Retrieved 12 January, 2021, from http://www.ug.ru/article/1029.

3. Belskikh, I. E., Boriskina, T. B., & Peskova, O. S. (2019). Features of the reflection of projects of Russian regional universities in the internet space. *Prospects of Science and Education, 2*(38), 465–477.
4. Belskikh, I. E., Peskova, O. S., Borodina, E. A., Kovazhenkov, M. A., & Yurova, O. V. (2018). Marketing and entrepreneurial potential of Russian universities as "locomotives" of regional economic growth. *Economics and Entrepreneurship, 7*(96), 752–754.
5. Gorbunov, A. P., Gorlova, E. B., & Maslennikova, N. V. (2016). *Theory and practice of education quality management in Russia.* Directmedia Publishing LLC.
6. Scientific Journal Fundamental Research. (2016). Development of business processes of an educational institution based on the application of IT infrastructure management standards. Retrieved 15 January, 2021, from https://fundamental-research.ru/ru/article/view?id=41081.
7. Tulchinsky, G. L. (2021). Digital transformation of education: challenges to higher education. Retrieved 10 January, 2021, from https://publications.hse.ru/articles/212164063.
8. Digital Transformation in the Education Sector—A Guide To Education Technology. Retrieved 17 January, 2021, from https://www.viewsonic.com/library/education/digital-transformation-education-sector-edtech/.
9. Robotic automation of processes. Reterieved 16 January, 2021, from http://www.tadviser.ru/index.php/Статья:RPA_Robotic_process_automation,_Роботизированная_автоматизация_процессов.
10. Sharon, H. (2021). Digital transformation of educational institutions. Retrieved 15 january, 2021, from https://vc.ru/flood/96913-cifrovaya-transformaciya-uchebnyh-zavedeniy.
11. Davletkireeva, L. Z., Novikova, T. B., & Kurzaeva, L. V. (2016). Development of business processes of an educational institution based on the application of IT infrastructure management standards. Retrieved 11 January, 2021, from https://www.fundamental-research.ru/ru/article/view?id=41081.
12. Melnik, O. (2019). Software robotics: without illusions. Retrieved 14 January, 2021, from https://www.it-world.ru/cionews/infrastructure/143831.html.
13. Uvarov, A. Yu. (2018). Education in the world of digital technologies: on the way to digital transformation. Moscow, Russia: ed. the house of the Higher School of Economics.
14. Chumachenko, M. A. (2020). "Quality of education" as a key factor in the marketing positioning of the university. *Modern Aspects of the Economy, 4*(272), 64–71.

Effectiveness of the Education System: Comparative Analysis of the Estimated Data Parameters

Irina A. Baranovskaia, Svetlana V. Derepasko, Nadezda P. Rudnikova, Irina A. Inyushina, and Ekaterina N. Khabaleva

Abstract The development of the education system and globalization of forms of interaction require the construction of general evaluation parameters of the data. First of all, this issue consists in studying the indicative base, which makes it possible to assess the effectiveness of the education system. The selected condition is due to the importance and significance of taking into account quantitative and qualitative changes that slow down the level of development of the education system in the Russian Federation and foreign countries. The main purpose of this study is to conduct a comparative analysis of the estimated parameters of the data in the area of determining the effectiveness of the education system. The result of the scientific research is the following aspects: the system of the Russian education is considered, the main institutions for monitoring the estimated parameters of data in the area of education are identified, and comparative analysis of the estimated parameters of data on the effectiveness of the education system of foreign countries and the Russian Federation is carried out. The research tools are the following methods: indicative analysis, grouping and generalization of data, comparison and consolidation of forms, sentences, identification, as well as strange differentiation of statistical information.

Keywords Education system · Parameters · Assessment · Comparative analysis · Efficiency · Indicators · Institutions · Monitoring

JEL Codes I20 · I21 · I129

I. A. Baranovskaia (✉) · S. V. Derepasko · N. P. Rudnikova
Orel State University, Orel, Russia
e-mail: 1278orel@mail.ru

I. A. Inyushina · E. N. Khabaleva
Orel State University of Economics and Trade, Orel, Russia

V. N. Ostrovskaya and A. V. Bogoviz (eds.), *Big Data in the GovTech System*, Studies in Big Data 110, https://doi.org/10.1007/978-3-031-04903-3_2

1 Introduction

Education as a system for the development of society in the XXI century has changed due to the allocation of human capital, skills, and competencies as fundamental factors of quality parity in the context of digitalization of the inter-country system of inter-connections. Education is associated with the level of economic development. If more investments are directed to the development of educational activities, then the themes of economic growth will be more significant in the medium term [11]. Despite this, the effectiveness of the education system depends on many factors that are calculated based on parameters that reflect the current state of the sector. The fragmentation of the estimated parameters of data between state institutions and countries doesn't allow forming a single image of an effective education system, which determines the significance and importance of the chosen research topic [16]. Let's consider the main difficulties in assessing the effectiveness of the education system.

Firstly, there is the fragmentation of indicators for assessing the effectiveness of the education system [12]. This problem is due to the establishment of priorities for the indicative base for the study of efficiency and the current level of development of the education system.

Secondly, the education system is a costly area of activity that produces results in a prolonged period [15]. The highlighted circumstance is associated with the prevailing subjective opinion in society states that education is an element of basic knowledge, which is relevant in the current period. Today, the globalization of the education system and the increased competition of human capital weaken the degree of significance of the statement above. The professional competition requires updating knowledge throughout life.

Thirdly, the authors noted the manifestation of the imbalance of inter-country human capital and the national education system [9]. This problem is caused by the specifics of building a national education system with the standards adopted in society. An example is a transition in 2003 of the Russian higher education system to the Bologna system. On the one hand, the population of the Russian Federation didn't personify a bachelor's degree with higher education for many years [10]. On the other hand, after the transition of the Russian higher education system to the Bologna system, the rudiments of a national community remained in its structure (for example, granting of candidate and doctor of sciences degrees).

In general, it can be noted that the problems highlighted above focus on the aspects of the importance of considering the parameters of evaluation data through the effectiveness of the education system. Based on this, the main purpose of writing a scientific article is a comparative analysis of the estimated parameters of the data, which determines the effectiveness of the education system. For achieving the stated goal, it's necessary to focus on solving the following tasks: to consider the Russian education system; to identify the main institutions for monitoring the evaluation parameters of data in the area of education; to carry out a comparative analysis of the estimated data parameters that determine the effectiveness of the education system of foreign countries and the Russian Federation.

2 Materials and Methods

The basis of the research is conditioned by the presence of articles by Russian and foreign authors in the area of studying the education system. The methodological toolkit for writing a scientific article is the method of exponential analysis [1], grouping and generalization of data, comparison and consolidation of forms, proposals, identification, as well as country differentiation of statistical information [8].

3 Discussion

So, let's consider the Russian education system. The education system in the Russian Federation includes a set of training programs based on the regulations of state standards in the area of education [19]. State standards in the area of education are implemented not only by state (municipal) institutions but also by independent organizations subordinate to the controlling and managing authorities. The Russian education system includes general and professional areas of education, which in turn are subdivided into basic and additional.

Table 1 shows the estimated parameters of the education sector data under the regulatory framework of the Russian Federation. The estimated data parameters presented in the national project "Education" can be noted as the most conditional.

A more structured system of evaluative data parameters is reflected in the Decrees of the Government of the Russian Federation in the area of implementation of the state program in the area of education, scientific and technological development of the Russian Federation, as well as monitoring of the education system. This aspect is expressed by the concretization of the estimated parameters of the data, due to the specialized analysis of the directions of the education system.

Thus, the state program "Development of Education" establishes an estimated data parameter showing the values of the rating indicators of the Russian Federation in comparison with the indicative number of foreign countries. According to this, the estimated data parameters focus on the place of the Russian Federation in:

- In the international study of the quality of reading and understanding of the text (PIRLS);
- In the international study of the quality of mathematical and natural science education (TIMSS) [18];
- In the international program for the assessment of educational achievement of students (PISA).

Table 1 Estimated parameters of education data under the regulatory framework of the Russian federation

National project "Education"	Decree of the Government of the Russian Federation of December 26, 2017 No. 1642 (as amended on October 7, 2021) "On approval of the state program of the Russian Federation 'Development of education'"
1. The entry of the Russian Federation into the top ten countries of the world in terms of the quality of general education	1. Russia's place in the international study of the quality of reading and text comprehension (PIRLS)
2. Formation of an effective system for identifying, supporting, and developing the abilities and talents of children and youth, based on the principles of justice, universality and aimed at self-determination, as well as the professional orientation of all students	2. Russia's place in the international study of the quality of mathematical and natural science education (TIMSS)
3. Creation of conditions for the upbringing of a harmoniously developed and socially responsible personality based on the spiritual and moral values of the peoples of the Russian Federation, historical, and national-cultural traditions	3. Russia's place in the international student assessment program (PISA)
4. Increase in the proportion of citizens engaged in volunteer activities or involved in the activities of volunteer organizations, up to 15%	4. The proportion of the number of graduates employed during the calendar year
	5. The number of Russian universities included in the top 100 world university rankings for at least 2 years in a row
	Added: 1.The number of people trained in online courses
	2. The number of graduates from colleges who have demonstrated a level of training that meets the standards of "WorldSkills"
Decree of the Government of the Russian Federation of March 29, 2019 No. 377 (as amended on September 11, 2021) "On approval of the state program of the Russian Federation. Scientific and technological development of the Russian Federation"	**Decree of the Government of the Russian Federation of August 5, 2013 No. 662 "On monitoring the education system"**
1. Expenses for scientific activities in the area of fundamental and applied scientific research (including the purchase of modern equipment)	1. Information about the organization carrying out educational activities in educational programs (form No. 1), form HE-1, form SVE-1, form GE-1
2. The number of foreign students studying in higher educational institutions in the Russian Federation	2. Information about the material, technical and information base, financial and economic activity (form No. 2), form HE-2, form SVE-2, form GE-2
3. Availability of higher education in the Russian Federation	

Source compiled by the authors

The results of the Russian Federation in the PIRLS study are determined by the following points: in 2001—16th place (528 points—an average value for 35 countries of 500 points), in 2006—1st place (565 points—an average value for 40 countries of 500 points), in 2011—2nd place (568 points—average value for 45 countries 500 points), in 2016—1st place (581 points—average value for 50 countries and 11 territories 500 points) [14]. In 2021, the study is being conducted in electronic format. The electronic format is based on the introduction to the study of the ePIRLS component, which determines how well students read, interpret, and evaluate information in an Internet-like environment. 66 countries are announced as participants. Participants explore web pages, answer questions, explain the relationship of phenomena, interpret, and integrate information.

The forms presented in Table 1 for general and vocational education regulate a single system of evaluative data parameters: basic information (for example, information about the organization, admission, the number of students and graduates, as well as the personnel of the organization), specific information on the presence of students or the results of admission (for example, the presence of foreign students, the number of students receiving scholarships, the results of admission in the current year), professional information (for example, whether teachers have a scientific degree, information on the level of education, experience and qualifications of employees), information on the material and technical base (for example, property of the organization, financial, and economic activities of educational institutions) [7].

The main publications on the estimated parameters of the education system data are the Russian statistical yearbook, preschool education (form No. 85-K), additional education for children (form No. 1-ADD), training of highly qualified personnel (form No. 1-SP), children's health-improving recreation (form No. 1-OL) [3]. The Higher School of Economics publishes a collection of indicators of education, including information on the activities of organizations that implement educational programs at various levels [5].

Within the framework of the study of identifying institutions for monitoring the estimated parameters of data in the area of education, the authors conclude that at the state level, the estimated parameters of data include a fairly large number of indicators that allow considering educational institutions, infrastructure, financial and economic activities, and the level of personnel training in absolute terms. At the same time, this estimated data parameter doesn't include an indicative base that assesses the effectiveness of the education system. However, it should be noted that today system of indicators has been created aimed at assessing higher education institutions in the Russian Federation, which comes down to the next four areas:

- Educational activities: (1) the average USE score of students admitted at the expense of budgetary funds or employers' funds, (2) the proportion of students at the university in all areas of educational activity;
- International activity: (1) the proportion of the number of foreign students;
- Scientific activity: (1) the number of publications indexed in the Scopus and Web of Science databases, (2) the growth rate of income received from R&D;

– Financial activity: (1) financial management index, (2) university revenues per one scientific and pedagogical worker, (3) university revenues per the number of students enrolled.

A separate line in the estimated data parameters is the indicator of the fulfilment of the quota for the employment of people with disabilities. Within the framework of the selected evaluation parameter of the data, it can be noted that the indicators used to determine only the current position of some areas of the education system, but they don't pay attention to the effectiveness of activities in this area. Except for this, without reflecting the effectiveness of the educational system, the emphasis is placed on the formation of statistical databases in the context of monitoring the sphere of education. Some indicators for assessing the education system are duplicated in different regulatory and program documents.

4 Results

The estimated parameters of these education systems in foreign countries are based on the availability of performance indicators and the availability of information on this sector. The educational systems of the United Kingdom, the United States and Germany were selected as the relevant set of scoring data. So, in the UK education system, the key monitoring institutions are the Office for National Statistics (which determines the index and rating base) [13], the Department for Education [2] (which forms the indicators of the general education system), and the Higher Education Statistics Agency [4] (research in the area of assessing data from the higher education system). Being part of the estimated data parameters, the Office for National Statistics monitors such indicators as labour productivity in education, the maximum level of teacher classification achieved in the region, as well as the number of students aged 16 and 24 years. The monitoring parameter of the Department for Education is due to the concentration on the development of the general education system through an assessment of the number of students, the average size of classes, the proportion of schools with an excess of the number of students, and the cost per student. Higher Education Statistics Agency compiles data statistics in the area of higher education assessment. The key indicators of the Higher Education Statistics Agency monitoring are the enrollment of students in higher education institutions at their place of residence and university, the number of teachers employed in teaching or research activities, a survey on the activities of graduates, who have completed their studies within the last 3 years.

In general, annual spending on the education system in the UK ranges from 4.8 to 6.5% of the level in the country's GDP (146 billion dollars in 2020). The UK high place in the Times Higher Education ranking of world universities, the leading positions of which are determined by the University of Oxford, University of Cambridge, UCL, Imperial College London, The University of Edinburgh (from 4th to 22nd place in the ranking) [17]. These universities have an impact not only on education but also on the economic development of the UK. So, 1 pound, invested in the University of Oxford, returns to the country's economy of 3.3 pounds.

The system of general education (junior and high school) in the UK is quite developed. Within the framework of the estimated parameterization of data, general education expresses social orientation and financial components that regulate the effectiveness of the education system. The financial set of estimated data parameters is reduced to the analysis of the revenue and expenditure side, including government grants, self-earned funds of educational institutions, expenses for teaching staff, rent, and other types of expenses (energy saving, training resources).

To assess the parameters of data based on the effectiveness of general education schools (primary and secondary schools), the coefficient of student progress (result) to the level of school funding (contribution) is used. Within the framework of similar assessment of data parameters for higher education institutions, the quality factor of the result is applied, based on the account of income from research activities and the number of diplomas issued.

In general, it can be noted that performance indicators are used as part of the estimated parameters of the UK education system data. An interesting indicator is the labour productivity indicator in education, which acts as an evaluative link, which makes it possible to conclude the contribution of each educational institution of education to the country's economy.

Spending on the education system in the United States varies annually from 5.6 to 6.2% of the country's GDP (1.2 trillion dollars in 2020) [6]. The main institutes for monitoring the scoring data are the U.S. Department of Education and the U.S. Department of Education's National Center for Education Statistics. The monitoring system of these institutions is based on studies of primary and secondary education (for example, indicators: the number of teachers, current expenditures on education, the share of income of state preschool and school institutions), higher education (for example, indicators: the number of students enrolled in higher educational institutions, the ratio of students full-time students to the total number of full-time staff and teachers), as well as the results and effectiveness of the education system (for example, indicators: labour productivity in education, the unemployment rate among persons aged 25 to 34 depending on the level of education, average annual earnings of permanent year-round workers from 25 to 35 years old, depending on the level of education and gender). In addition, the monitoring system uses a general evaluative data parameter aimed at studying education expenditures in comparison to the level of GDP and a comparative analysis of international data in the area of education, based on rankings of the best universities (according to the Times Higher Education ranking, the Massachusetts Institute of Technology Institute (MIT), Stanford University, Harvard University, University of Chicago).

Within the framework of assessing the effectiveness of education in the United States, the most developed data parameters are used to analyze the higher education system. First of all, this performance assessment is aimed at reducing the expenditure side of the higher education system through the following tasks:

- Optimization of business processes in higher education institutions;

- Remote interaction of students and teachers in the process of research activities;
- Complementary cooperation of educational institutions based on joint research and general interuniversity academic programs.

As for indicators of the effectiveness of the education system, an evaluation parameter is used, aimed at highlighting the productivity of education through the number of students per 1 teacher (tutor) and the number of costs for their provision. The estimated parameters of the data in the German education system are built a little differently. Expenditures on the German education system annually range from 4.0 to 4.4% of the country's GDP (172.4 billion dollars in 2020). The Federal Ministry of Education and Research is responsible for the monitoring of data estimates in the German education system.

The dedicated evaluative data parameters are used throughout the European Union as part of the study of the system for evaluating the effectiveness of the education sector. While studying the effectiveness of the German educational system, indicators are used to assess (1) the proportion of people who haven't completed their studies, (2) the level of student failure in several disciplines (for example, in natural science subjects), (3) the employment rate of specialists who have recently finished training.

In general, within the framework of the countries presented above, there is a commonality of the estimated data parameters in the education system, taking into account national characteristics. The use of a single assessment system by all countries allows ensuring the comparability of statistical information and harmonizes and unifies the indicators used as part of a single criterion. In addition to assessing the effectiveness of the education system in foreign countries, the indicator "labour productivity in education" is used. This feature can be taken into account in the context of monitoring the education system of the Russian Federation. Today, the category of labour productivity in education is not enshrined in the legislative acts of the Russian Federation. The key assessment database is the study of the share of government spending on the education system as a percentage of GDP. It's advisable to consider labour productivity in education as a system of complementary indicators covering the key elements of the activities of educational organizations. At the same time, labour productivity in education must have a quantitative characteristic. As a proposal, formula aimed at using the costs of providing the education system to the production of products, which is manifested through the number of students, assessed through the result of examination sessions (with an average final score in the range of 4–5) can be applied:

$$p_x = \frac{z}{c}, \tag{1}$$

where p_x is labour productivity in education, z is the total government spending on the education system, and, finally, c is the number of students assessed through the results of examination sessions.

5 Conclusion

The conducted research on the effectiveness of the education system within the framework of a comparative analysis of the parameters of the estimated data made it possible to draw the following conclusions.

1. Identification of the main monitoring institutions made it possible to form a system of evaluative parameters in the area of education and to identify key problems that hinder the unification and harmonization of data with the values of indicators of international statistics. The estimated data parameters used in the Russian Federation don't include an indicative base to assess the effectiveness of the education system. Some indicators for assessing the education system are duplicated in regulatory and program documents. Another estimated parameter of the data doesn't allow reflecting the real situation in the development of the education system.
2. As part of a comparative analysis of the estimated parameters of data on the effectiveness of the education system based on foreign countries (the UK, Germany and the United States), key indicators were established that make it possible to fully form an idea of the current situation in the development of this area. But, in addition to using indicators of the effectiveness of the education system, an estimated data parameter is used, including labour productivity in education. This experience can be applied in monitoring the Russian education system.

References

1. Boyarskikh, E. V., & Molozhavenko, V. L. (2020). The effectiveness of the education quality management system in the municipal autonomous educational institution. *Internauka, 20–1*(149), 67–68.
2. Department for Education. Retrieved 14 October, 2021, from https://www.gov.uk.
3. Federal State Statistics Service of the Russian Federation. Education. Retrieved 15 10, 2021, from https://rosstat.gov.ru.
4. Higher Education Statistics Agency. Retrieved 18 October, 2021, from https://www.gov.uk.
5. Higher School of Economics. Education indicators (2021). Retrieved 18 October, 2021, from https://www.hse.ru.
6. IMF Data. Retrieved 14 October, 2021, from https://data.imf.org.
7. Lyapina, I., Sotnikova, E., Lebedeva, O., Makarova, T., & Skvortsova, N. (2019). Smart technologies: Perspectives of usage in higher education. *International Journal of Educational Management, 33*(3), 454–461.
8. Mammadov, R., & Çimen, I. (2019). Optimizing teacher quality based on student performance: A data envelopment analysis on PISA and TALIS. *International Journal of Instruction, 12*(4), 767–788.
9. Mikk, J., Krips, H., Säälik, Ü., & Kalk, K. (2016). Relationships between student perception of teacher-student relations and PISA results in mathematics and science. *International Journal of Science & Mathematics Education, 14*(8), 1437–1454.
10. Mitrofanova, E. P. (2020). Efficiency and quality of personnel training in the system of secondary vocational education using the tools of the national qualifications system. *Modern Education: Topical Issues and Innovations, 2*, 22–27.

11. Naumov, D. I., & Nikitina, I. Yu. (2014). Social efficiency of the system of higher professional education: theoretical aspect. Bulletin of Berdyansk State Pedagogical University, Series 2: History. *Philosophy, Political Science, Sociology, Economy, and Culturology, 4*(82), 84–89.
12. Nurgalieva, S. A., Mailybaeva, G. S., Asylova, R. O., & Utegulov, D. E. (2019). The effectiveness of the education system in Kazakhstan: PISA assessment. *Bulletin of West Kazakhstan State University, 2*(74), 19–32.
13. Office for National Statistics. Retrieved 1 October, 2021, from https://www.ons.gov.uk.
14. PIRLS (International study of the quality of reading and text comprehension). Retrieved 10 September, 2021, from https://fioco.ru.
15. Shmatova, A. P., & Mezentseva, E. M. (2020). The effectiveness of the use of the point-rating system in the area of education. *Student Bulletin, 24–4*(122), 101–102.
16. Stroeva, O., Zviagintceva, Y., Tokmakova, E., Petrukhina, E., & Polyakova, O. (2019). Application of remote technologies in education. *International Journal of Educational Management, 33*(3), 503–510.
17. The Times Higher Education. Retrieved 18 October, 2021, from https://www.timeshighereducation.com.
18. TIMSS (International study of the quality of mathematics and science education). Retrieved 15 October, 2021, from https://fioco.ru.
19. Vasilyeva, N. V. (2020). The influence of new ways of thinking and informatization of the education system on the effectiveness of learning. *Professional Education, 1*(39), 26–29.

Supplementary Legal Material

20. Decree of the Government of the Russian Federation of December 26, 2017 No. 1642 (as amended on October 7, 2021) "On approval of the state program of the Russian Federation 'Development of education". Retrieved 15 October, 2021, from https://base.garant.ru/71848426/.
21. Decree of the Government of the Russian Federation of March 29, 2019 No. 377 (as amended on September 11, 2021) "On approval of the state program of the Russian Federation. Scientific and technological development of the Russian Federation". Retrieved 15 October, 2021, from https://base.garant.ru/72216664/.
22. Decree of the Government of the Russian Federation of August 5, 2013 No. 662 "On monitoring the education system". Retrieved 15 October, 2021, from https://base.garant.ru/70429494/.

Modern Educational Platforms for Distance Training on Lean Production

Zhanna V. Smirnova, Elena A. Chelnokova, Zhanna V. Chaykina, Evgeny A. Semakhin, and Denis S. Kostylev

Abstract This article discusses the modern educational platform for distance learning on Lean manufacturing. The author disclosed the content of the developed electronic courses on Lean manufacturing. The purpose of the study was determined - this is the formation of professional competencies of students in professional activity. The main tasks were set: development and implementation of model for training a future teacher using the concept of «Lean production», including the modernization of the content, forms, methods and means of teaching through a distance learning format, the development of e-learning course on «Lean production». The main advantages of modern distance-learning platforms were considered. A model for organizing the teacher learning process in the context of the formation of professional competencies in lean technologies has been developed. The author studied theoretical material about the use of electronic educational platforms for teaching lean production to students. The conclusion that the developed distance course on «Lean technologies» is one of the most relevant platforms for the transfer of educational information for students, increasing the quality and accessibility of education, was justified.

Keywords Distance learning · Lean production · Educational platforms · Economy · Efficiency

JEL Codes R11 · R12 · R58 · Q13 · Q18

Z. V. Smirnova (✉) · E. A. Chelnokova · Z. V. Chaykina · E. A. Semakhin
Minin Nizhny Novgorod State Pedagogical University, Nizhny Novgorod, Russia
e-mail: z.v.smirnova@mininuniver.ru

E. A. Chelnokova
e-mail: chelnokova_ea@mininuniver.ru

D. S. Kostylev
Institute of Food Technology and Design, Nizhny Novgorod State University of Engineering and Economics, Nizhny Novgorod, Russia

V. N. Ostrovskaya and A. V. Bogoviz (eds.), *Big Data in the GovTech System*, Studies in Big Data 110, https://doi.org/10.1007/978-3-031-04903-3_3

1 Introduction

We are increasingly faced with the policy of competitiveness of educational services with the development of modern education system.

The effectiveness of organization of educational process will largely depend on the conditions of educational activity. One of the tools for improving the educational process is the information and communication environment of the university, the development of distance learning format for students.

At the moment, there is a variety of remote platforms on which it is possible to conduct a training format on the organization of lectures, practical classes and independent work. The choice of a platform largely depends on their ability to conduct educational activities [1].

In this situation, when educational institutions are moving into a distance learning format, online services are becoming an urgent issue of using distance educational platforms. There is a demand for the purchase of online resources related to remote work activities, the organization of project activities, the creation of electronic training courses.

2 Methodology

As part of the study, we considered the issue of using distance learning in the process of studying a course on Lean production.

The aim of the study is the formation of professional competencies of students in professional activities. The main objectives of the project are the development and implementation of a model for training a future teacher using the concept of «Lean production», including the modernization of the content, forms, methods and means of teaching through a distance learning format, the development of e-learning course on «Lean production» [2].

The developed model includes target, meaningful and effective components that allow, in general, to assess the level of formation of professional competencies.

The teachers of Minin University have developed the model for the formation of professional competencies in the implementation of the concept of teaching a subject area, and the e-learning course on «Lean production» has been developed for university students on the basis of the meaningful form of the concept to improve the quality of training future teachers.

In the course of the research, the author studied theoretical material on the use of electronic educational platforms for teaching lean production to students. The problem of using lean technologies in educational institutions has been studied, the issue of introducing professional competencies into the professional activities of future teachers has been considered in the works of I.A. Volkova [9]. A.G. Chernov's information [3] on the organization of the activities of educational institutions using lean production technologies has been studied.

Thus, the set tasks of the study reflect the process of building a model for the formation of professional competencies of future teachers in the field of lean technologies at Minin Nizhny Novgorod State Pedagogical University (Minin University). An assessment of the effectiveness of the development of an electronic distance learning course on lean production has been carried out.

The study involved the teachers of the Faculty of Management and Social and Technical Services of Minin University (12 people) and students of the first and second year of study in the amount of 34 people.

3 Results

Minin University began to train educators with lean competencies. Minin University has been preparing teachers with lean competencies within the framework of the project «The end-to-end flow of forming a lean personality of a teacher» since September 2021 [5].

Future teachers will explore lean technologies that are already widely used in production along with subjects from their professional field. This is the 5S concept, the Paretto principle, the «Five whys» technique as a problem-solving tool, the «Red label» method for eliminating unnecessary and non-value-adding time management technologies.

Lean technology is about effective teacher time management. Teachers will have more time to work with students and improve the educational process, rather than shifting papers, preparing reports and looking for tools.

At the moment, lean technologies have been introduced as additional training. Their inclusion in the curriculum for all students in the future is being considered [6].

In the process of research, we have developed the model for training a future teacher using the concept of «Lean production», which includes the modernization of the content, forms, methods and means of teaching through a distance learning format at the university (Fig. 1).

The future teacher develops new professional competencies as part of improving the teacher training system using lean technologies in the learning process: he quickly adapts to the educational process, forms the ability to independently learn new methods of optimizing lean production, forms the ability to organize and develop methodological support for the learning process using lean production technologies.

The construction of a system of the educational process according to this model is based on the fact that the content of theoretical training in all courses of study includes the formation of competencies in lean production. The volume of curriculum hours at the current stage allows the formation of lean production competencies with the use of additional education courses [8].

We have developed a distance course of additional education using elements of distance learning «Lean technologies» to ensure the quality of training for the formation of professional competencies using lean technologies [4].

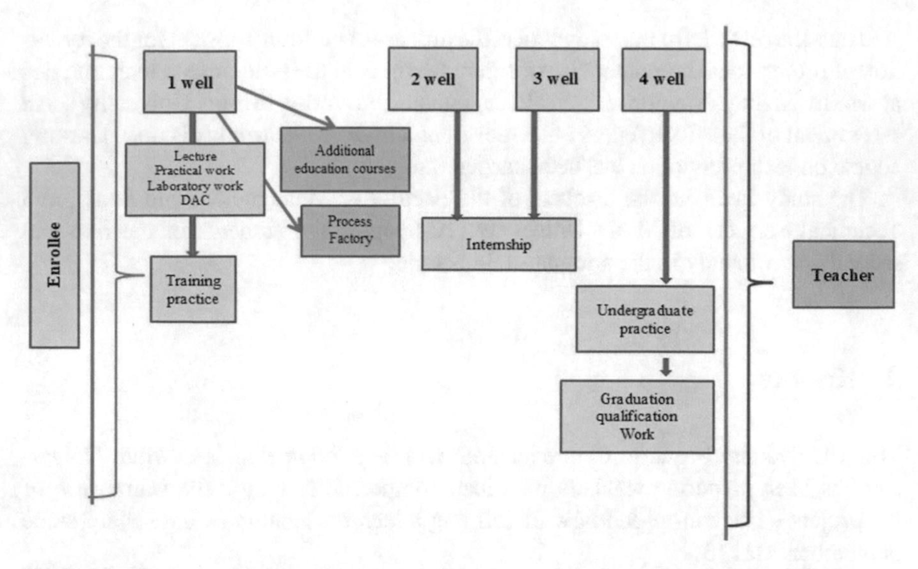

Fig. 1 Model of organization of teacher training process. *Source* Developed and compiled by the authors

Today, lean production technologies are the most effective method of organizing educational activities at the university.

The content of Lean production contains many tools for the process, management system and culture of the organization.

The content of the distance course includes lecture material, practical work, topics of independent work, as well as control of knowledge on the passed material (test tasks), using educational platforms in the Moodle system [7].

Figure 2 shows the fragment of the developed course of additional education for students of Minin University.

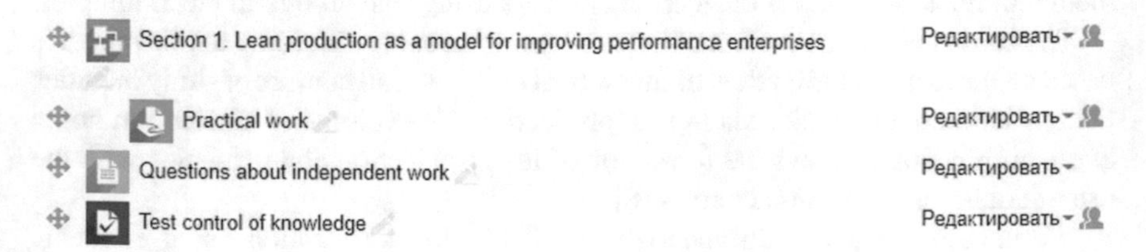

Fig. 2 Distance course «Lean technologies». *Source* Developed and compiled by the authors

Distance learning courses are designed for 72 h, in the process of studying, the student must know: the patterns of organizational work in the unit and organization for the implementation of lean technologies; be able to: set the tasks of tactical and strategic development in organizations using TPS, critically analyze the practice of solving them; own: traditional and modern management technologies for solving management problems [10].

Based on the material covered, students develop professional competencies in lean technologies, which they can successfully apply in their future professional activities.

4 Conclusion

Thus, Minin University plans to expand the number of students in the formation of lean competencies on the basis of the developed model of organizing the process of training a teacher and taking advanced training courses according to the developed distance electronic course «Lean technologies».

The introduction of professional competencies in lean technologies has a positive effect on the production process of future teachers.

The developed distance course on «Lean technologies» is one of the most relevant platforms for the transfer of educational information for students which increases the quality and accessibility of education.

References

1. Bazhenov G.E., Dyachkova A.V., Problems of implementing the concept of lean production at Russian enterprises // Business. Education. Right. Bulletin of the Volgograd Institute of Business. (2016). August. *No., 3*(36), 14–20.
2. Burmasheva, E. P. (2016). *Use of lean production instruments in designing the educational process // Integration of Education., 20*(1), 105–111. https://doi.org/10.15507/1991-9468.082.020.201601.105-111
3. Chernov A.G. Theoretical aspects of the application of lean technologies in education. – Nizhny Novgorod: NP PC In 2 vols. - T.I: «Logos», 2021. 80 p.
4. Efimov V.V. Fundamentals of lean production: tutorial. Ulyanovsk: ULSTU, 2011.
5. Garina, E. P., Romanovskaya, E. V., Andryashina, N. S., Kuznetsov, V. P., & Shpilevskaya, E. V. (2020). *Organizational and Economic Foundations of the Management of the Investment Programs at the Stage of Their Implementation Lecture Notes in Networks and Systems, 91*, 163–169.
6. Lizunkov V.G., Morozova M.V., Zakharova A.A., Malushko E.Yu. On the issue of criteria for the effectiveness of interaction between educational organizations and enterprises of the real sector of the economy in the conditions of advanced development territories // Vestnik of Minin University, 2021 Volume 9, No. 1.
7. Sedykh E.P. System of normative legal support of project management in education. Vestnik of Minin University. 2019 T. 7. No. 1 (26). P.
8. Stomma M.V. Lean Learning // Human Resources Handbook. 2009. No. 10.

9. Volkova I.A. Sectoral features of the implementation of the thrift system // Business. Education. Right. Bulletin of the Volgograd Institute of Business. 2016. No. 3(36). pp. 21–25.
10. Womack James P., Jones Daniel T. Lean Manufacturing: How to get rid of losses and make your company prosperous / Translated from English. 2nd ed. M.: Alpina Business Book, 2005. 473 p. (Series «Management models of leading corporations»).

The Use of Digital Technologies in the Implementation of the Meta-Subject Approach as a Trend in the Development of the International Educational Environment

Mariia V. Mukhina, Olga V. Katkova, Sergey E. Revunov, Zhanna V. Smirnova, and Tatyana N. Tsapina

Abstract Digital technologies today are not just a requirement of time and an obligatory component of the international educational environment; it is a powerful tool that allows the development of personal competencies in the conditions of remote learning, in the implementation of comprehensive and complex educational tasks. One of such tasks is the implementation of a meta-subject approach at all levels of the educational process. The purpose of this paper is to reveal the methodology of using digital technologies in the implementation of the meta-subject approach as a means of developing a personal holistic imaginative perception of the world. In the research conducted, the implementation of the meta-subject approach is shown on the example of the trainees' technological training. The educational standards and other regulatory documents emphasize the importance of the meta-subject approach and the importance of using digital technologies for its implementation. The authors have developed and described a methodology for implementing the meta-subject approach by information means during technology lessons, which includes three basic components: the technology conceptual basis; procedural characteristics; resource support. An example of the developed information and computer tasks for one of the learning stages is given. This technique will help the teacher plan the lesson and achieve the necessary level of development of meta-subject results. The advantages of the technique are the possibility of using it as a diagnostic of the level of development of meta-subject results, as well as the possibility of using it in remote learning.

M. V. Mukhina (✉) · O. V. Katkova · S. E. Revunov · Z. V. Smirnova
Minin Nizhny Novgorod State Pedagogical University, Nizhny Novgorod, Russia
e-mail: mariyamuhina@yandex.ru

O. V. Katkova
e-mail: katkova.ov@yandex.ru

S. E. Revunov
e-mail: revunov@inbox.ru

Z. V. Smirnova
e-mail: z.v.smirnova@mininuniver.ru

T. N. Tsapina
Lobachevsky State University of Nizhny Novgorod, Nizhny Novgorod, Russia

V. N. Ostrovskaya and A. V. Bogoviz (eds.), *Big Data in the GovTech System*,
Studies in Big Data 110, https://doi.org/10.1007/978-3-031-04903-3_4

Keywords Digital technologies · International educational environment · Meta-subject approach · Information means · Meta-subject approach implementation method · Trainees

JEL Codes I23 · I25

1 Introduction

The development of the contemporary international educational environment is not possible without the use of digital technologies that ensure the transition of the subject of the educational process to a new qualitative level [1].

Currently, the choice of digital technologies that can be used to improve the efficiency of the educational process is very wide: from digital educational platforms (Open Platform of the National Education, International educational platforms, electronic platforms of Russian universities, etc.) to individual digital tools and services (virtual laboratories, mental maps, online boards, services for creating interactive exercises, etc.) [2, 3].

The use of digital technologies in the educational process allows forming not only the professional competencies required for each individual related to a specific activity, but also the digital competence that is relevant today, allowing to work effectively with information of any complexity, databases, navigate the digital space and quickly master new forms and methods of work in a changing professional environment [4, 5].

For successful integration into the international educational and professional environment, each student must have a willingness to join the changeable world with a high degree of uncertainty. Therefore, the desire to form a personal holistic imaginative perception of the world is one of the main tasks of the educational process at all levels of education [6]. This task is most effectively solved by the meta-subject approach.

The meta-subject approach significance is determined by the main regulatory documents of the educational sphere. According to the Federal State Educational Standard, the meta-subject competencies are included in the list of learning outcomes that students are to master [7].

The meta-subject approach in school education is an explicit expression of the integration processes taking place today in science and in the life of society. These ties play an important role in improving the practical and scientific-theoretical training of students, an essential feature of which is the mastery of the generalized nature of cognitive activity by schoolchildren.

The meta-subject approach in domestic education has been developed in the works of A.V. Khutorskoy, N.V. Gromyko and Yu.V. Gromyko, A.A. Musina, later becoming one of the main guidelines for the creation of Federal State Educational Standards [8].

In the works of Yuri V. Gromyko, the meta-subject content of education is interpreted as an activity that does not relate to any particular subject, while ensuring the educational process when teaching any academic subject. I.e., the "principle of meta-subject" is the basis in teaching general means, techniques, methods of mental activity of students, can and should be used when working with any educational material, regardless of the academic subject.

In the works of A.V. Khutorskoy, the meaning of the meta-subject character of education is clearly traced. In his opinion, it lies in the fact that the main essence of education is to identify, develop and realize the inner potential of a person not only in relation to oneself and one's inner world, but also to search for ties between the person's inner and outer worlds, which is achieved through activities that relate to the solid foundations of the world and man [8].

One of the required conditions for the implementation of the meta-subject approach is the use of information tools. The State Program of the Russian Federation "Education Development" for 2013–2020 indicates the need to create an environment using information and communication learning tools in the education system.

Information learning tools are electronic means of storing, processing and transmitting educational information using electronic devices.

Information and communication technologies (ICT) make it possible to achieve those trainee skills and abilities that are interpreted in the Federal State Educational Standard (FSES) [7]. Therefore, the purpose of the research is to study the theoretical foundations of the meta-subject development and to develop a methodology for conducting lessons using information tools aimed at the formation of trainees' meta-subject results.

The practical significance lies in the fact that the developed methodology has a practice-oriented nature and can be used in practical activities at technology lessons at school in a remote mode.

2 Methodology

The methodological basis of the research is the requirement of the international educational system to prepare a person for life in the modern world. The student's possession of the generalized nature of cognitive activity, the ability to a holistic imaginative perception of the world is a criterion of readiness for professional activity. The development of meta-subject knowledge, skills and abilities successfully contributes to solving of these tasks. The basis for the implementation of the meta-subject approach is the system-activity approach, which is aimed at the active, independent, cognitive activity of the trainee. The system-activity approach provides not only the mastering of the training program, but also the development of meta-subject results that are necessary for students to survive in a dynamically changing world. The use of information technologies in the implementation of the meta-subject approach contributes to the development of students' digital skills. The integration of these conceptual provisions formed the basis for the development of a methodology for the implementation of the meta-subject approach using information means during technology lessons.

During the research, the following methods were used: analysis of scientific literature, periodical articles, thematic publications, pedagogical experiments, results processing.

3 Results

The research results have revealed that the meta-subject approach is an extensive concept that is actively used in modern school education. It allows students to achieve a holistic imaginative perception of the world, moving away from subject-based learning, contributes to the better formation of various concepts within individual subjects, groups and systems, thus contributing to shaping of a personality in demand by contemporary society.

An effective tool for achieving meta-subject results is the use of modern information technologies in the educational process.

As the Technology learning subject requires the study of a number of theoretical and practical materials, which is almost impossible to present to students in practice, visualization of the studied material is necessary, which is successfully achieved using ICT. Information tools, in addition, make it possible to implement remote learning methods, which is relevant for the modern educational system [9–11].

As part of the work on the research topic, we have developed a methodology for implementing a meta-subject approach using information means during technology lessons. The educational methodology is understood as an ordered system of actions, the implementation of which leads to the guaranteed achievement of training goals. The description of the developed methodology contains the following three main parts: methodology conceptual basis; procedural characteristics; resource support.

Here’s a brief description of the methodology.

The first component called Methodology Conceptual Basis contains:

- FSES requirements to personality development,
- Methodological approach,
- Remote training mode,
- Informatization of education,
- Bloom’s Taxonomy

The developed methodology is based on the requirements of the Federal State Educational Standard (FSES) for personality development, for the formation of universal (meta-subject) educational actions. It is necessary to take into account the up-to-date requirements for the need to switch on to remote learning and to informatization of education.

The general principle of the ICT competence development according to the Federal State Educational Standard is the formation of specific technological skills in the course of their meaningful application. The implementation of the meta-subject approach in the educational environment using ICT is carried out as a result of the development of universal educational actions of trainees, such as cognitive, regulatory, communicative ones. Thus, the use of information tools, especially in technology lessons, is a prerequisite for the implementation of a meta-subject approach in teaching schoolchildren.

In the process of the methodology development work, we have relied on the B. Bloom's teaching goals taxonomy, which includes six categories of goals with corresponding sub-goals: knowledge (specific material, terminology, facts, definitions, criteria, etc.) → understanding (explanation, interpretation, extrapolation) → application → analysis (relationships, principles of construction) → synthesis (development of a plan and a possible system of actions, obtaining a system of abstract relations) → evaluation (judgment based on available data, judgment based on external criteria).

The second component called Procedural characteristics of methodology contains:

- Technology subject program,
- Calendar-thematic plan,
- Level-deep knowledge system,
- Electronic assignment options,
- Methodological recommendations on technology use,
- Technology advantages

When developing this component, the Technology subject program was analyzed, which allows identifying opportunities for the use of a meta-subject approach implementing information means during technology lessons [12]. A calendar and thematic plan was also developed, sections and topics of lessons were identified for which it is advisable to develop information and computer assignments.

The main constituent part of this component is the creation of information and computer tasks, examples of which have been developed for each level of the methodology. Information and computer tasks (ICT) were developed on the bases of the platforms: Diary.ru https://dnevnik.ru/. Russian Electronic School https://resh.edu.ru/, LearningApps.org https://learningapps.org/about.php/. The platforms offer different interfaces [2], so the tasks have included various elements: video materials, quizzes, tests, simulated exercises and tasks on the material mastered to reinforce the knowledge, etc. The materials of the lessons have been adapted to the architecture of the platforms used.

Here's an example of the information-computer tasks developed by us for one of the stages.

Stage one of the logic reasoning development is knowledge. For example, it can be a retelling of educational material, accompanied by a list of facts, definitions, concepts, description of simple processes. A similar list can be built using a web service "EdWordle.net."

The trainees retell a number of terms and concepts from the delivered educational material, from which a digital word cloud is formed. The main criterion by which a teacher can judge the improvement of a thematic dictionary is the "word cloud" replenishment.

Please see below a lesson fragment which uses this internet service:

Lesson topic: Properties of structural materials, class 6.

Lesson type: integrated (technology, computer science).

Lesson goal:

(1) for the teacher: to introduce students to the technology of wood harvesting, the basic physical and mechanical properties of wood and reinforce knowledge using the "Word Cloud" service (http://www.edwordle.net);
(2) for the trainees: to study the physical and mechanical properties of wood, to create a cloud of words on this topic.

Planned results:

(1) subject-wise: the trainees know the physical and mechanical properties of wood, are able to determine the density and humidity of wood;
(2) meta-subject-wise: the trainees show the ability to determine the degree of success of their work, evaluate the work of classmates.

The technology teacher's activity: delivers a lesson on the topic "Properties of structural materials", uses a presentation with illustrations and videos for clarity.

The computer science teacher's activity (at the stage of applying the criteria for compiling the cloud and evaluating the method): organizes the work of Internet services, prints the finished cloud, takes a screenshot, inserts it into the class blog or presentation.

Students' activities: listen and record the lesson material, answer the technology teacher's questions, work with the EdWordle application: choose "Create now"; make a list of words or sentences; form a cloud (change the font, color, shape and location of words at will); save a picture ("Save image") (Fig. 1).

Such a process of constructing a text in a digital format and its visualization contributes to the acquisition of new knowledge not by memorizing it in a ready-made form, but in the course of assignment with the help of information tools. This method meets the needs of modern children, which means that it reduces the time

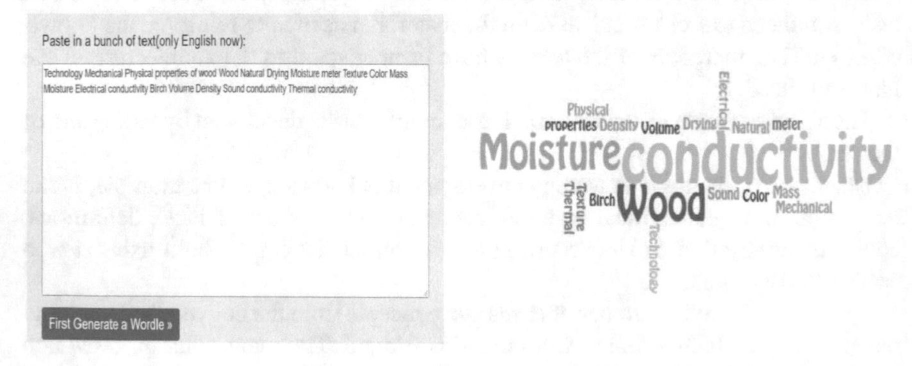

Fig. 1 Word Cloud on the Properties of structural materials topic. *Source* Developed and compiled by the authors

to master new concepts and increases the frequency of their application in everyday life.

The third component of the developed methodology is the resource support. Its composition is:

- Computer, computer class available,
- Steady internet connection,
- Internet educational resources
- HR resource

This component includes the need for a computer, a computer class – a computer or laptop in the basic configuration will be quite enough for our methodology. To perform some tasks and implement the remote training format, the device must have a stable Internet connection. High speed is not required for a comfortable work.

To implement most of the stages of the methodology, the teacher needs to have only a basic level of IT competence. For designing more complex tasks of levels 4,5,6, for example, designing a room or apartment using platform www.planner5d.com, one can resort to the help of a computer science teacher. Thus, the integrated lesson format is used, which is also one of the ways to implement the meta-subject approach.

4 Conclusion

Thus, the developed methodology will help the teacher plan the lesson and achieve the necessary level of development of meta-subject results. Having determined the topic, the goal of the lesson and the planned results, the teacher selects from the developed information-communication tasks (ICTs) the ones of a certain level, fills them up with the lesson material, and applies them in the learning process. When using the developed ICTs, it becomes possible to simulate any type of lesson. The advantages of this methodology are as follows:

(1) Allows implementing a meta-subject approach, providing a level-by-level mastering of knowledge by the trainees, which makes it possible to build an individual learning trajectory.
(2) Due to its flexibility, it can be used in remote learning conditions. Since the developed tasks of the methodology are based on an information and computer format, they can be performed by the trainees remotely if they have a computer and an Internet access.
(3) It can be used in other disciplines with the adjustment of the content of tasks.
(4) It is convenient by the simplicity of diagnostics. The tasks in the methodology are designed in such a way that the indicator of mastering a certain level of development of meta-subject results is the very fact that the trainee has completed the tasks. Thus, the teacher receives a tool that will help not only

bring the development of students' meta-subject knowledge to the required level, but also diagnose the result obtained.

Despite the growing trend in the entire international educational environment to introduce digitalization into the education system, the degree of use of information technology by teachers remains at a low level.

To activate the information and educational environment, the teacher needs to analyze various electronic educational platforms, master these and place lessons and assignments with various interesting, interactive materials. This will increase the efficiency of the educational process, will help facilitate the transition to remote learning and will contribute to the formation of the trainee's digital competencies.

References

1. Iordache, C., Marien, I., & Baelden, D. (2017). Developing digital skills and competences: A quick-scan analysis of 13 digital literacy models. *Italian Journal of Sociology of Education, 9*(1), 6–30. DOI: https://doi.org/10.14658/pupj-ijse-2017-1-2. Retrieved 9 September, 2021, from https://www.researchgate.net/publication/315690154_Developing_Digital_Skills_and_Competences_A_Quick-Scan_Analysis_of_13_Digital_Literacy_Models.
2. Kullaslahti, J., Ruhalahti, S., & Brauer, S. (2019). Professional development of digital competences: Standardized frameworks supporting evolving digital badging practices. *Journal of Siberian Federal University. Mathematics and Physics, 12*, 175–186. Doi: https://doi.org/10.17516/1997-1370-0387. (In Eng.). Retrieved 18, September, 2021, from http://elib.sfu-kras.ru/bitstream/handle/2311/109543/Kullaslahti.pdf;jsessionid=294AB7EB4DC6198265ED8C43E01A048E?sequence=1.
3. Loshkareva, E., Luksha, P., Ninenko, I., Smagin, I., & Sudakov, D. (2017). Skills of the future. How to Thrive in the new complex world, Moscow, WorldSkills Russia, p. 98. (In Russ.). Retrieved 9 September, 2021, from http://spkurdyumov.ru/future/navyki-budushhego-chto-nuzhno-znat-i-umet-v-novom-slozhnom-mire/.
4. Malushko, E. Yu., Lizunkov, & V. G. (2020). The e-education system as a tool to improve the competitiveness of a specialist in the digital economy. Vestnik of Minin University. Vol. 8, No. 2. Doi: https://doi.org/10.26795/2307-1281-2020-8-2-310.26795/2307-1281-2020-8-2-3. Retrieved 19 September, 2021, from https://vestnik.mininuniver.ru/jour/article/view/1083.
5. Samerkhanova, E. K., & Balakin, M. A. (2020). Preparation of managers of professional educational programs to work in the digital environment of the university. Vestnik of Minin University. Vol. 8, No. 2. Doi: https://doi.org/10.26795/2307-1281-2020-8-2-4. Retrieved 5 October, 2021, from https://vestnik.mininuniver.ru/jour/article/view/1084.
6. Katkova, O., Mukhina, M., Chaykina, Zh., Smirnova, Zh., & Tsapina, T. (2021). Analysis of electronic educational resources for distance learning. In the collection: Education and City: Education and Quality of Living in the City. The Third Annual International Symposium. Moscow. 5032. Retrieved 06 October, 2021, from https://www.researchgate.net/publication/349918326_Analysis_of_electronic_educational_resources_for_distance_learning.
7. Stankevich, O. V. (2017). Metasubject approach in modern education in the context of the implementation of the Federal State Educational. Young Scientist. No. 50. pp. 271–274. [Electronic resource]. Retrieved 22 September, 2021, from https://moluch.ru/archive/184/47158/.
8. Khutorskoy, A. V. (2017). Didactics: a textbook for universities. Third generation standard. SPb.: Peter. – p. 720. Retrieved 06 October, 2021, from http://khutorskoy.ru/books/2017/Khutorskoy_A.V._Didaktika/.
9. Akhmetzhanova, G. V., & Abieva, D. R. (2019). Peculiarities of the teacher's activity in the conditions of distance learning. *Azimuth of Scientific Research: Pedagogy and Psychology,*

3(28), 28–31. DOI: https://doi.org/10.26140/anip-2019-0803-0005. Retrieved 08 September, 2021, from https://cyberleninka.ru/article/n/osobennosti-deyatelnosti-pedagoga-v-usloviyah-distantsionnogo-obucheniya.
10. Gryaznova, E. et al. (2019). Distance pedagogy: problems of conceptualization and definition. *Azimuth of Scientific Research: Pedagogy and Psychology, 8*(4)(29), 63–65. DOI: https://doi.org/10.26140/anip-2019-0804-0015. Retrieved 19 September, 2021, from https://cyberleninka.ru/article/n/distantsionnaya-pedagogika-problemy-kontseptualizatsii-i-opredeleniya.
11. Mukhina, M. V., & Mukhina, E. S. (2019). Distance learning as a modern trend in the development of society. *Innovative Economy: Prospects for Development and Improvement, 6*(40), 57–64. Retrieved 10 October, 2021, from https://www.elibrary.ru/item.asp?id=41228326.
12. Tishchenko, A. T. (2017). Technology: work program: grades 5–9. NV Sinitsa. – M.: Ventana-Graf. – p. 158. Retrieved 22 September, 2021, from https://rosuchebnik.ru/material/tekhnologiya-5-9-klassy-rabochaya-programma-tischenko/.

Digital Skills Shaping as a Factor of Sustainable Development of Higher Education

Mariia V. Mukhina, Nikolay A. Barkhatov, Olga V. Katkova, Zhanna V. Smirnova, and Zhanna V. Chaykina

Abstract The trend of introducing digital technologies into all spheres of human activity sets the task of the digital skills development for the present-day educational sphere the training of a future specialist. The paper presents the study results of the undergraduate students digital skills shaping issues and the capabilities of digitalization of electronic educational courses in the disciplines of university training based on the example of the educational environment of the Minin University of Nizhny Novgorod. The purpose of the study: to review readiness of the students' digital skills shaping by means of the university educational disciplines. The relevance of the study is due to the increasing demand for digital skills of future specialists from employers. During the study process, the courses of Minin University have been analyzed for the available content aimed at the digital skills shaping within these disciplines. The data on the provision of electronic educational courses with information technology elements are presented. Additional opportunities for the digital skills shaping have been identified. Examples of tools that are feasible to be implemented in the electronic educational courses. The Moodle platform has been analyzed for the presence of mechanisms for involving students in the game process. It has been shown that a university teacher can create a high-quality course with gaming basic elements in the electronic environment of the university without special IT training. Recommendations have been developed for the modernization of electronic educational courses in accordance with the requirements of the up-to-date digital reality.

M. V. Mukhina (✉) · N. A. Barkhatov · O. V. Katkova · Z. V. Smirnova · Z. V. Chaykina
Minin Nizhny Novgorod State Pedagogical University, Nizhny Novgorod, Russia
e-mail: mariyamuhina@yandex.ru

N. A. Barkhatov
e-mail: nbarkhatov@inbox.ru

O. V. Katkova
e-mail: katkova.ov@yandex.ru

Z. V. Smirnova
e-mail: z.v.smirnova@mininuniver.ru

Z. V. Chaykina
e-mail: jannachaykina@mail.ru

V. N. Ostrovskaya and A. V. Bogoviz (eds.), *Big Data in the GovTech System*,
Studies in Big Data 110, https://doi.org/10.1007/978-3-031-04903-3_5

Keywords Digital skills · Digital technologies · University training · Electronic educational courses · Moodle electronic platform

JEL Codes I23 · I25

1 Introduction

Education is one of the important factors determining the priority directions of the social development [1]. The educational process should be efficient and mobile, responding to all trends of social development. Presently we are witnessing a rapid change in the educational environment towards the population digital skills development. The introduction of digital technologies in all spheres of human activity is convincing that humanity is entering a new high-tech digital level of development [2].

The digital skills issue is discussed in the international expert field both in authoritative publications of international organizations and leading analytical centers (World Economic Forum [3], Worldskills [4], UN [5]) and at the level of governments of different countries [6, 7]. Russia also faces a movement towards both the population digital skills development and the digital economy development as a whole. By Order No. 1632-r of the Government of the Russian Federation dated July 28, 2017, the RF Digital Economy Program is included in the list of the main RF strategic development trends for 2017–2030. According to this program, the share of the population with digital skills shall be 40% [8]. The implementation of this program in the Russian Federation has begun several years ago and will continue in the future. A large number of currently conducted scientific theoretical and practical research in this area testifies to the constant search for new ways and methods to improve the digital literacy of the population [9, 10].

Digital skills mastered by a modern specialist, is a prerequisite for a successful professional activity [11]. Digital skills are understood as "well-established, automated behavioral models based on knowledge and skills in the use of digital devices, communication applications and networks to access and manage information" [12]. The demand for digital skills among modern employers is constantly growing and the skill shaping is, first of all, assigned to the university training of bachelors.

This study has examined how modern disciplines taught to university students are aimed at the digital skills shaping. The study has been conducted based on the example of the analysis of the disciplines of the electronic educational environment of the Minin University of Nizhny Novgorod.

Thus, the purpose of the work is to investigate the readiness of the students' digital skills shaping by means of the university educational disciplines.

2 Methodology

The study of the digital skills development issue in the process of student training has been conducted on the basis of the Minin University of Nizhny Novgorod. The electronic educational courses in various training areas have been analyzed. To carry out the analysis, criteria were developed to give an objective assessment of the provision of courses with elements of information technology. The methodological basis of the research is the theoretical analysis method, the regulatory documentation analysis, the data generalization method, methods of quantitative and qualitative processing of the results obtained.

3 Results

The Minin University has been training the students in various areas, both pedagogical and non-pedagogical, which are in demand in the modern labor market. Education at the Minin University is based on the Moodle platform, ensuring the availability of online and off-line education for students [5, 13]. The active use of e-courses was provoked by the aggravated epidemiological situation, which led to a forced transition to a remote education [14–16]. The remote mode immersion revealed the problem of the students' digital skills mastering and showed that possession of digital skills is a necessary component of students' training [17–19].

Within the framework of the research conducted, 15 courses of social and economic and natural science disciplines of the Minin University have been analyzed (https://ya.mininuniver.ru). The following research criteria have been put forward:

1. availability of meaningful content of the discipline related to the information technology;
2. presentation of methodological support related to the use of digital technologies in electronic educational and methodological complexes;
3. presence of special literature in the discipline work programs, the content of which is related to the digital skills shaping.

The analysis showed that in disciplines that are not directly related to computer technologies, there is no meaningful content that would disclose content specifically aimed at using digital technologies in relation to the discipline. At the same time, the inclusion of this content in the electronic educational course gives the student an opportunity to get acquainted with the role and influence of digital resources on a certain scientific direction. The potential provided by digital technologies is huge and the task of the teacher is to show these opportunities to the student in relation to the educational discipline.

According to the second criterion, some information technologies that develop digital skills are presented in the work programs and in the electronic educational and methodological complexes (EEMC). The most widely used are Zoom conferences, tests, forums, chats, Web 2 (Table 1).

Table 1 Information technologies and resources used

No	Information technologies and resources	Percent of use
1	Use of telecommunication technologies (network resources) for webinars, on-line conferences: ZOOM and other information resources	100
2	Chat, forum	95
3	Test	95
4	Hyperlinks to open full-text editions of educational and scientific literature, periodicals and other electronic educational resources	90
5	Hyperlinks to open full-text EEMC author(s) editions	15
6	There are links to illustrations and borrowed material (textbooks, manuals), hyperlinks to presentations, videos, audio materials and other educational resources, including EEMC resources as well	65
7	Use of additional software, including network training programs	15
8	Use of cloud technologies and Web 2.0 resources to organize joint activities	52

Source Developed and compiled by the authors

As can be seen from the table, teachers do use digital technologies in the methodological support of the electronic educational courses, but not well enough. Resources such as the use of virtual laboratories, the ability to conduct experiments in a safe environment, including those that are not feasible in a normal class room, for example, measurements of radioactive radiation, studying changes in electric current in different conditions or the principles of engine operation "from the inside", etc., will provide a modern student with a new quality of education [20].

According to the third criterion, the absence of specialized literature in the studied disciplines, the content of which is associated with the digital skills shaping, was revealed. The inclusion of specialized literature will open up the opportunity for students to gain knowledge independently, orient them in large amounts of information and provide an opportunity to choose the necessary sources of knowledge for themselves.

The study results described have showed that there are additional opportunities for the undergraduate students' digital skills development, which are not fully used in electronic educational courses and disciplines of the university training. In the course of the work carried out, it has been revealed that the digital environment of the Internet space contains many tools that can be used as additional tools for the development of students' digital skills. Here are some of these: Padlet (https://ru.padlet.com/)—this service allows the student to place his/her work on the blackboard, and the teacher to comment and evaluate everyone; Miro (https://miro.com/), Lucidspark (https://lucidspark.com/ru—virtual boards for the implementation of joint ideas; Mentimeter (https://www.mentimeter.com/) is a tool for voting and creating interactive presentations, providing instant feedback from the audience, etc. [21].

Gamification of educational courses has great opportunities for the digital skills development. Gamification is widely used in digital learning platforms and is beginning to penetrate into the classical education system. The Moodle platform used at the Minin University to host e-courses has a significant potential for including this tool in educational courses. Such features of the platform as encouragement with points, badges, certificates, motivation by competition through flow distribution and access restriction, as well as the capabilities of the rating system as an element of a competitive resource, chat functions, forum, individual messages and operational contact with the teacher ensure the optimal trajectory of individual learning, the deadline function can be a key tool that will increase efficiency and reveal qualities that are not available in other ways [22, 23]. The education gamification process can be provided with the help of other educational platforms: Khan Academy https://ru.khanacademy.org/, Cousera https://ru.coursera.org/, Udemy https://www.udemy.com/, Memrise https://www.memrise.com/ru/, Yousician https://yousician.com/, etc.

Most university teachers can create a high-quality course with basic gamification elements in the electronic environment of the university without special IT training [24–26].

4 Conclusion

The demand for digital skills and the employees with digital skills increases with the continuous evolution of the digital economy, and in the future it will grow even more, therefore, the undergraduate students' digital skills development is a prerequisite for their entry into the modern professional activity. Disciplines and university training courses should ensure such development. At the present stage, modernization of educational courses is required in accordance with the requirements of the up-to-date digital reality. In order to accomplish this, the following will be required:

- to supplement the content of disciplines and courses with the educational content reflecting the use of digital technologies in science and practice of the human activity;
- to introduce additional tools and services to the methodological component of the disciplines taught at the university for the development of digital skills in the classroom and in the course of independent student work;
- to include in the list of references special sources, the content of which is related to the discipline digital skills shaping.

Therefore, the digital literacy improvement is not a new-fashioned trend at all, but a vital process for the contemporary society, which shall be provided by the education system at all its stages.

References

1. Samerkhanova, E. K., & Balakin, M. A. (2020). Preparation of managers of professional educational programs to work in the digital environment of the university. *Vestnik of Minin University 8*(2). https://doi.org/10.26795/2307-1281-2020-8-2-4. Retrieved 05 October, 2021, from https://vestnik.mininuniver.ru/jour/article/view/1084.
2. Katkova, O. V., Mukhina, M. V., & Gureeva, E. P. (2021). Problems of the formation of digital skills of university students. Science of Krasnoyarsk. T. 10. No. 3–3. pp. 76–80. Retrieved 19 September, 2021, from https://www.elibrary.ru/item.asp?id=44833265.
3. Breene, K. (2016). The 10 Countries Best Prepared for the New Digital Economy, (In Eng.). Retrieved 08 September, 2021, from https://www.weforum.org/agenda/2016/07/countries-best-prepared-for-the-new-digital-economy/.
4. Kullaslahti, J., Ruhalahti, S., & Brauer, S. (2019). Professional development of digital competences: Standardized frameworks supporting evolving digital badging practices. *Journal of Siberian Federal University. Mathematics and Physics, 12*, 175–186. Doi: 10.17516 / 1997–1370–0387. (In Eng.). Retrieved 18 September, 2021, from http://journal.sfu-kras.ru/en/article/109543.
5. The five largest online learning platforms. The largest educational platforms are collected in the review of RIAMO [Electronic resource]. Retrieved 22 September, 2020, from https://riamo.ru/article/231962/5-krupnejshih-platform-onlajn-obucheniya.xl?mTitle=&mDesc.
6. Bayuan, H. Joraev, & Pakhmutova, M. A. (2019). Features of the electronic educational environment of a modern university. DOI: https://doi.org/10.33910/herzenpsyconf-2019-2-37. Retrieved 18 September, 2021, from https://www.researchgate.net/publication/338161540_Features_of_the_electronic_educational_environment_of_the_modern_university.
7. Retrieved 19 September, 2021, from https://www.umj.ru/jour/article/view/1368/0.
8. Gryaznova, E. et al. (2019). Distance pedagogy: Problems of conceptualization and definition. *Azimuth of Scientific Research: Pedagogy and Psychology, 8*(4)(29), 63–65. DOI: https://doi.org/10.26140/anip-2019-0804-0015. Retrieved 19 September, 2021, from https://cyberleninka.ru/article/n/distantsionnaya-pedagogika-problemy-kontseptualizatsii-i-opredeleniya.
9. Digital Economy Program. (2017). Order of the Government of the Russian Federation of July 28, 2017, No. 1632-r, (In Russian.). Retrieved 19 September, 2021, from http://publication.pravo.gov.ru/Document/View/0001201708030001.
10. Loshkareva, E., Luksha, P., Ninenko, I., Smagin, I., & Sudakov, D. (2017). Skills of the future. How to Thrive in the New Complex World, Moscow, WorldSkills Russia, p. 98. (In Russ.). Retrieved 9 September, 2021, from http://spkurdyumov.ru/future/navyki-budushhego-chto-nuzhno-znat-i-umet-v-novom-slozhnom-mire/.
11. Katkalo, V. S. et al. (2018). Teaching digital skills: Global challenges and best practices. Analytical report for the III International Conference «More than Learning: How to Develop Digital Skills». M.: ANO DPO «Corporate University of Sberbank», p. 122. Retrieved 09 September, 2021, from https://sberuniversity.ru/upload/iblock/2f8/Analytical_report_digital_skills_web_demo.pdf.
12. Scheerder, A. J., Van Deursen, J. A., & Van Dijk, G. M. (2019). Taking advantage of the internet: A qualitative analysis to explain why educational background is decisive in gaining positive outcomes Poetics Available online 16 December 2019, 101426 https://doi.org/10.1016/j.poetic.2019.101426. Retrieved 06 October, 2021, from https://www.sciencedirect.com/science/article/abs/pii/S0304422X19300889.
13. Malushko, E. Yu., & Lizunkov, V. G. (2020). The e-education system as a tool to improve the competitiveness of a specialist in the digital economy. *Vestnik of Minin University, 8*(2). https://doi.org/10.26795/2307-1281-2020-8-2-3. Retrieved 19 September, 2021, from https://vestnik.mininuniver.ru/jour/article/view/1083.

14. Bulaeva, M. N., Vaganova, O. I., Koldina, M. I., Lapshova, A. V., & Khizhnyi, A. V. (2018). Preparation of bachelors of professional training using MOODLE. *Advances in Intelligent Systems and Computing, 622,* 406–411. DOI: https://doi.org/10.1007/978-3-319-75383-6_52. Retrieved 09 September, 2021, from https://www.researchgate.net/publication/323568357_Preparation_of_Bachelors_of_Professional_Training_Using_MOODLE.
15. Dmitriev, Ya. V., Alyabin, I. A., Brovko, E. I., Dvinina, S. Yu., & Demyanova, O. V. (2021). Fostering University Students' Digital Skills: De Jure vs De Facto. *University Management: Practice and Analysis, 25*(2), 59–79. Doi: https://doi.org/10.15826/umpa.2021.02.015. (In Russ.).
16. United Nations E-Government Survey. (2018). (In Russ.). Retrieved 18 September, 2021, from https://publicadministration.un.org/publications/content/PDFs/UN%20E-Government%20Survey%202018%20Russian.pdf.
17. Gruzdeva, M. L., Golubeva, O. V., Mukhina, M. V., Chaikina, Z. V., & Cherney, O. T. (2021). Information technologies to support the study of disciplines. *Studies in Systems, Decision and Control, 314,* 1163–1171. Retrieved 18 September, 2021, from https://link.springer.com/chapter/https://doi.org/10.1007/978-3-030-56433-9_122.
18. Mukhina, M. V., & Mukhina, E. S. (2019). Distance learning as a modern trend in the development of society. *Innovative Economy: Prospects for Development and Improvement, 6*(40), 57–64. Retrieved 10 October, 2021, from https://www.elibrary.ru/item.asp?id=41228326.
19. Vaganova, O. I., Chelnokova, E. A., Smirnova, Z. V., Mukhina, M. V., & Ponomareva, E. (2020). Organizing E-Learning using. *Cloud TechnologiesInternational Journal of Advanced Trends in Computer Science and Engineering, 9*(4), 4844–4848, 94. Retrieved 22 September 2021, from http://www.warse.org/IJATCSE/static/pdf/file/ijatcse94942020.pdf.
20. Iordache, C., Marien, I., & Baelden, D. (2017). Developing digital skills and competences: A quick-scan analysis of 13 digital literacy models. *Italian Journal of Sociology of Education, 9*(1), pp. 6–30. Doi: https://doi.org/10.14658/pupj-ijse-2017-1-2. Retrieved 09 September, 2021, from https://www.researchgate.net/publication/315690154_Developing_Digital_Skills_and_Competences_A_Quick-Scan_Analysis_of_13_Digital_Literacy_Models.
21. Srba, I., Savic, M., Bielikova, M., Ivanovic, M., & Pautasso, C. (2019). Employing community question answering for online discussions in university courses: Students' perspective Computers & Education, Vol. 135, pp. 75–90. Doi: https://doi.org/10.1016/j.compedu.2019.02.017. Retrieved 19 September, 221, from https://www.sciencedirect.com/science/article/abs/pii/S0360131519300466.
22. Akhmetzhanova, G. V., & Abieva, D. R. (2019). Peculiarities of the teacher's activity in the conditions of distance learning. *Azimuth of Scientific Research: Pedagogy and Psychology, 3*(28), 28–31. Doi: https://doi.org/10.26140/anip-2019-0803-0005. Retrieved 08 September, 2021, from https://cyberleninka.ru/article/n/osobennosti-deyatelnosti-pedagoga-v-usloviyah-distantsionnogo-obucheniya.
23. Vaganova, O. I., Aleshugina, E. A., & Maksimova, K. A. (2019). Design of electronic training courses Azimuth of Scientific Research: Pedagogy and Psychology, *8*(3)(28), 57–59. Doi: https://doi.org/10.26140/anip-2019-0803-0013. Retrieved 19 September, 2021, from https://cyberleninka.ru/article/n/proektirovanie-elektronnyh-uchebnyh-kursov.
24. Katkhanova, Yu. F., Avetisyan, D. D., & Kirilov, D. Yu. (2016). "Cloudy" portal in the electronic educational environment. Research: From theory to practice: Materials of the IX Intern. Scientific-practical conf. Cheboksary: Central nervous system "Interactive plus", pp. 98–102. ISSN 2413–3957. Doi: https://doi.org/10.21661/r-112401. Retrieved 06 October, 2021, from https://interactive-plus.ru/ru/article/112401/discussion_platform.

25. Katkova, O., Mukhina, M., Chaykina, Zh., Smirnova, Zh., & Tsapina, T. (2021). Analysis of electronic educational resources for distance learning. In the collection: Education and City: Education and Quality of Living in the City. The Third Annual International Symposium. Moscow, p. 5032. Retrieved 06 October, 2021, from https://www.researchgate.net/publication/349918326_Analysis_of_electronic_educational_resources_for_distance_learning.
26. Vaganova, O. I., Smirnova, Z. V., Mukhina, M. V., Kutepova, L. I., & Chernysheva, T. L. (2017). The organization of the test control of students' knowledge in a virtual learning environment Moodle. *Journal of Entrepreneurship Education, 20*(3). Issue: 3. [Electronic resource]. Retrieved 22 September, 2020, from https://www.abacademies.org/articles/the-organization-of-the-test-control-of-students-knowledge-in-a-virtual-learning-environment-moodle-6919.html.

Digital Technologies in the Teacher's Professional Activities

Zhanna V. Smirnova, Elena A. Chelnokova, Mariia V. Mukhina, Olga T. Cherney, and Elena P. Kozlova

Abstract The utilization of digital technologies in the educational activities of the university teacher is considered in the research. The author identified the goal of the study, which is to theoretically and practically substantiate the readiness of university teachers to use information and communication technologies and to analyze how to improve the effectiveness of the learning process with the use of these technologies. This study is considered through monitoring the formation of the educator competences in the area of information and communication technologies. Information and communication technologies used by university students in the process of distance learning are considered by the author in the work. A questionnaire survey of Minin University teachers was carried out in the process of monitoring teachers on their readiness to use computer technologies in educational activities. The questionnaire was attended by university teachers with different work experience and age category. Based on the study, a conclusion was made about the effectiveness of the utilization of digital technologies in the educational activities of a teacher, about the influence of information technologies on the quality of educational training of university students.

Keywords Professional activity · Digital technologies · Teacher · Learning process · Information and communication technologies

JEL Codes R11 · R12 · R 58 · Q13 · Q18

Z. V. Smirnova (✉) · E. A. Chelnokova · M. V. Mukhina · O. T. Cherney · E. P. Kozlova
Minin Nizhny Novgorod State Pedagogical University, Nizhny Novgorod, Russia
e-mail: z.v.smirnova@mininuniver.ru

E. A. Chelnokova
e-mail: chelnokova_ea@mininuniver.ru

M. V. Mukhina
e-mail: mariyamuhina@yandex.ru

O. T. Cherney
e-mail: fiolet1975@mail.ru

E. P. Kozlova
e-mail: elka-a89@mail.ru

V. N. Ostrovskaya and A. V. Bogoviz (eds.), *Big Data in the GovTech System*,
Studies in Big Data 110, https://doi.org/10.1007/978-3-031-04903-3_6

1 Introduction

Today, one of the important areas of development of the world economy is the informatization of education using up-to-date information technology.

Constant informatization of society and the use of modern information and communication technologies in the learning process are among the outstanding trends of the twenty-first century. They have a significant advantage in the transfer of cognitive information to the student. In such conditions, digitalization of education is aimed at the formation and improvement of the development of a modern intellectual society in a certain environment of informatization.

The process of improving forms of educational activity and its content with adding computer-based teaching methods requires solving a number of problems in the teacher's educational activities.

Thus, the quality of the educational process as a whole will depend on the improvement of the informatization of the system of education.

2 Methodology

The final aim of the study is a theoretical and practical substantiation of the readiness of university teachers to use information and communication technologies and analysis of improving the effectiveness of the learning process using these technologies. We consider this study as monitoring the formation of the competencies of university educators in the field of information and communication technologies and the advanced training of teachers in accordance with the demands of the digitalization of the economy [1].

On the basis of this goal, we identified the research tasks: to conduct a study in modern information technologies, to analyze the use of modern technologies in the learning process at the university, to assess the readiness of university teachers to use digital technologies, to analyze the main problems of using information and communication technologies at the university.

The research process was carried out by the method of questioning 34 university teachers, the respondents were asked to evaluate their activities with the work of digital technologies in the learning process. The study was carried out at Minin Nizhny Novgorod State Pedagogical University (Minin University) [2].

The university is one of the leading universities in the use of digital technologies, the development of informatization makes it possible to conduct educational activities in the context of information technologies. Nevertheless, not everyone of the faculty membership have mastered professional skill in the field of information and communication technologies. The task of the university is to achieve the highest indicators in the field of digitalization of the learning process, including professional development of teachers [3].

3 Results

In the process of research, we examined modern information technologies that are used in the learning process at Minin University.

The access to educational resources in the format of E-Learning (Moodle) is organized at Minin Nizhny Novgorod State Pedagogical University. E-learning is being introduced by the university, both in full-time and part-time studies. The organization of e-learning is carried out using distance educational technologies.

Distance learning is a synthesis of interactive self-study and intensive consulting support. Thus, e-learning can be considered one of the tools of distance education.

Modern educational technologies used at Minin University include: research teaching methods, a «portfolio» assessment system, lecture and seminar systems, problem-based learning, project technologies, a collective learning system, modular learning technologies, etc.

These technologies are successfully applied in pedagogical work with the use of the electronic educational environment Moodle.

Thus, today, one of the requirements for university teacher is the acquisition of skills to work on computers and in electronic educational resources. The teacher must have the appropriate competence in the field of information technology.

In the present day, one of the problems of the university is the limited competence in the use of information technologies by educators [4].

A questionnaire survey of Minin University teachers was carried out in the process of monitoring teachers for their readiness to use computer technologies in educational activities. The questionnaire was attended by university teachers with different work experience and age category. The result of the survey showed that 28% of teachers are hesitant to use digital technologies or do not use them at all, 19% of teachers use digital technologies and the majority of teachers 53% actively use the Internet and they have no difficulties in working with a computer, they are interested in new applications and software resources and they actively use social networks (Fig. 1).

After analyzing the main problems of using information and communication technologies by the university, according to teachers, it was concluded that the level of competence in information and communication technologies is low. The electronic resources used at the university are constantly updated, this contributes to the low productivity of the teacher's work with the resource, there is a constant need to learn certain skills by creating specific information in electronic courses [6].

According to the survey results (Fig. 2), we see that most of the teachers do not have enough advanced training courses in information and communication technologies (ICT), 49% of people want to improve their ICT competencies, 28% study and work with programs independently, 23% do not want to take courses at all.

Improving the level of qualifications of a teacher largely depends on the level of preparation for classes. The teacher needs to keep up with the development of information technology, to master new formats for lecture studies, laboratory work and independent work using innovative technologies.

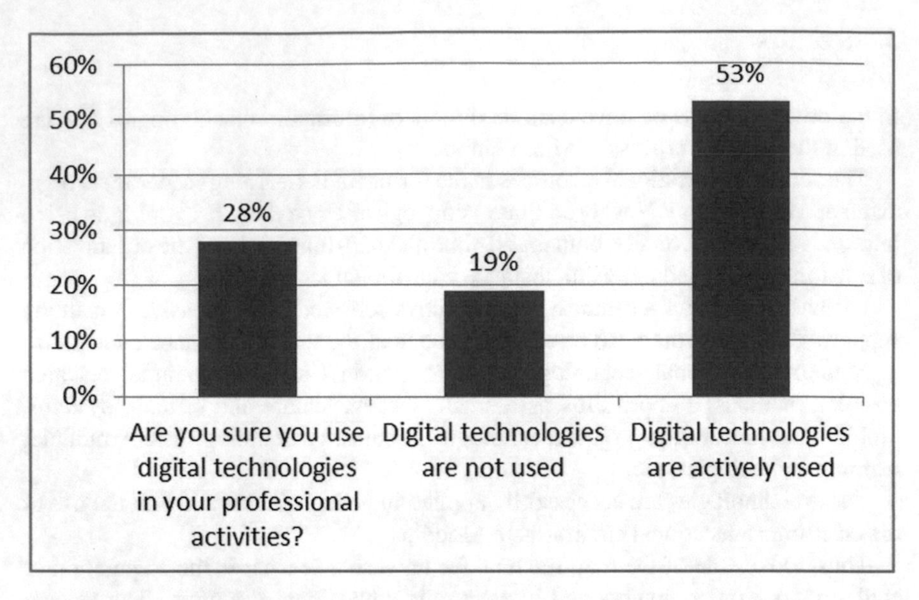

Fig. 1 Readiness of teachers to use computer technologies in educational activities. *Source* Developed and compiled by the authors

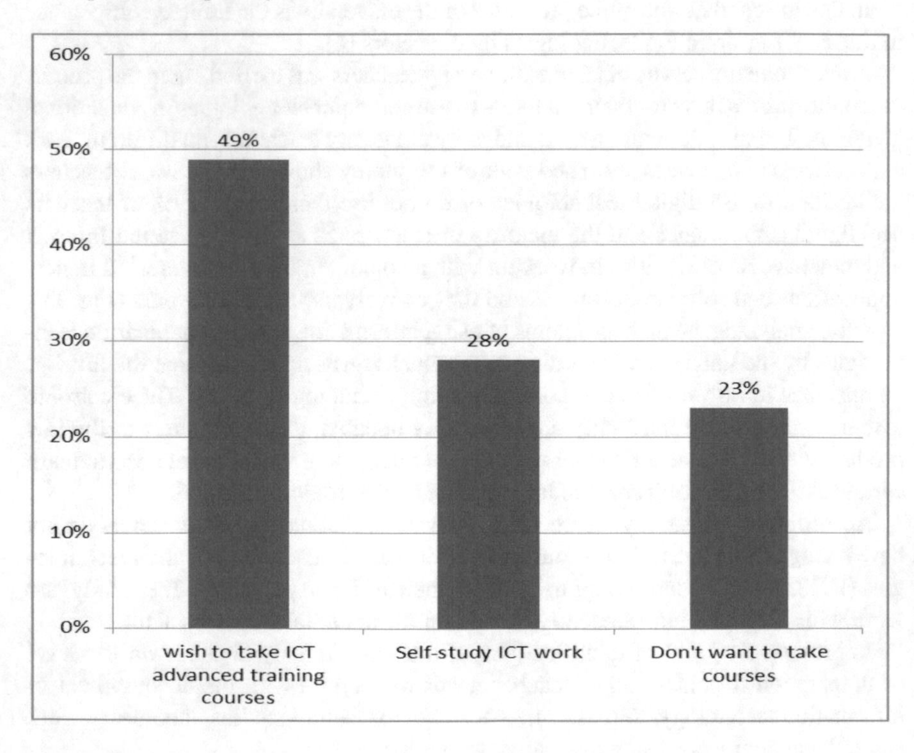

Fig. 2 Professional development courses for teachers: «for» and «against». *Source* Developed and compiled by the authors

The use of digital technologies in teacher's activities leads only to positive results in mastering educational information by students [7].

According to students, classes with the use of information technologies are becoming more interesting, accessible and modern, in addition, knowledge acquires consistency and complex application, increases the focus of the interests of university students.

In connection with the exploitation of modern techniques, it becomes necessary to assess the effectiveness of their use by mentors. Evaluation of the effectiveness of the use of latest educational technologies by educators in the educational process is carried out according to the following criteria:

Criterion № 1. Level of proficiency in modern educational technologies and methods.

Criterion № 2. Effectiveness of the use of modern educational technologies.

Criterion № 3. Personal investment to the improving of educational quality through the implementation of modern educational technologies.

At Minin University, despite the fact that some of the teachers are not actively involved in working with innovative digital methods of teaching disciplines, there is also a positive result of the active use of information and communication technologies in educational activities: the material and technical base in the classrooms of the university is being updated, modern equipment is being installed with digital technologies, the atmospheric environment of classrooms, in which non-traditional forms of classes with the use of ICT are held, is being improved [5, 9].

The university administration organizes advanced training courses in the field of ICT for university employees in a distance learning format, which is convenient for the teacher. New training programs in disciplines are being purchased, the level of digitalization is being improved in the documentary support of the teacher, and work with documents is being simplified [8, 10, 11].

Thus, the use of information and communication technologies in the professional experience of educator gives rise to the upgrading of the quality of students' training, increases interest in the process of cognition, performance of work and the study of the material of disciplines in general.

4 Conclusion

The study of the application of digital technologies in the professional experience of university teacher is one of the key points of information society development. The need to develop distance learning in the educational activities of the university is determined by the solution of a number of tasks: analysis of up-to-date information and communication technologies in the educational space of the university, improving the education system based on the use of ICT, introducing distance educational programs into educational activities, e-learning.

Analysis of the research data identified that the use of ICT in the professional activity of a teacher is having an effective impact on the quality of student learning outcomes and enhances the worldview interest in cognitive activity in general.

References

1. Ablakotov, A. A. (2016). The use of ICT on technology lessons is a factor in increasing the activation of cognitive activity. A.A. Ablakotov, V.V. Epaneshnikov. – M.: The whole world, p. 600.
2. Afanasyev, A. N. (2013). Theoretical aspects of the application of modern computing technologies in education. A. N. Afanasyev, R. P. Ivanov. – SPb.: Ripol Classic, p. 320.
3. Garina, E. P., Kuznetsov, V. P., Egorova, A. O., Romanovskaya, E. V., & Garin, A. P. (2017). Practice in the application of the production system tools at the enterprise during mastering of new products. Contributions to Economics. № 9783319606958. pp. 105–112.
4. Garina, E. P., Romanovskaya, E. V., Andryashina, N. S., Kuznetsov, V. P., & Shpilevskaya, E. V. (2020). *Organizational and Economic Foundations of the Management of the Investment Programs at the Stage of Their Implementation Lecture Notes in Networks and Systems, 91*, 163–169.
5. Kuznetsov, V. P., Romanovskaya, E. V., Egorova, A. O., Andryashina, N. S., & Kozlova, E. P. (2018). Approaches to developing a new product in the car building industry. *Advances in Intelligent Systems and Computing. T, 622, 494–501.*
6. Lizunkov, V. G., Morozova, M. V., Zakharova, A. A., & Malushko E. Yu. (2021). On the issue of criteria for the effectiveness of interaction between educational organizations and enterprises of the real sector of the economy in the conditions of advanced development territories. Vestnik of Minin University, Vol. 9, No. 1.
7. Polat, E. S. (2015). New pedagogical and information technologies in the education system. E.S. Polat et al. M. AST-press. – 2015. – 36, pp. 54–58.
8. Sedykh, E. P. (2019). System of normative legal support of project management in education. *Vestnik of Minin University. T, 7*(1)(26).
9. Smirnova, Z. V., Kuznetsova, E. A., Koldina, M. I., Dyudyakova, S. V., & Smirnov, A. B. (2020). *Organization of an Inclusive Educational Environment in a Professional Educational Institution Lecture Notes in Networks and Systems, 73*, 1065–1072.
10. Smirnova, Z. V., Vaganova, O. I., Chanchina, A. V., Koldina, M. I., & Kutepov, M. M. (2020). Development of research activity of future economists in the university. In collection: Lecture Notes in Networks and Systems. Growth Poles of the Global Economy: Emergence, Changes and Future Perspectives. Plekhanov Russian University of Economics. Luxembourg, pp. 371–379.
11. Zakharova, I. G. (2015). Information technologies in education: textbook. I.G. Zakharova. – M.: Academy, p. 198.

Distance Learning in Higher Education: Technologies of the Moodle Electronic Environment

Zhanna V. Smirnova, Marina L. Gruzdeva, Elena A. Chelnokova, Mariia V. Mukhina, and Svetlana N. Kuznetsova

Abstract This article examines the applying of distance learning techniques in the educational activities of the university. The author shows the content of distance technologies in the electronic educational environment of the university. The relevance of the research is that distance learning technologies allow using new forms of learning in educational activities, regardless of the location of the student, and the ability to constantly update the content of the academic process. In the paper, the study was carried out at the Faculty of Management and Social and Technical Services of Minin University. To implement these conditions, information and communication technologies (ICT) are used at the university, courses of disciplines are developed in the Moodle system. The structure and content of electronic courses are considered. The author conducted a survey of students on the benefits of distance learning in improving the quality of educational services. The study analyzed the obtained data on the effectiveness of distance learning, applied methods, ways, types, tools, directions and models of distance learning.

Keywords Distance learning · Technology · Electronic environment · Moodle

JEL Codes R11 · R12 · R58 · Q13

Z. V. Smirnova (✉) · M. L. Gruzdeva · E. A. Chelnokova · M. V. Mukhina · S. N. Kuznetsova
Minin Nizhny Novgorod State Pedagogical University, Nizhny Novgorod, Russia
e-mail: z.v.smirnova@mininuniver.ru

M. L. Gruzdeva
e-mail: gruzdeva_ml@mininuniver.ru

E. A. Chelnokova
e-mail: chelnokova_ea@mininuniver.ru

S. N. Kuznetsova
e-mail: dens@52.ru

V. N. Ostrovskaya and A. V. Bogoviz (eds.), *Big Data in the GovTech System*,
Studies in Big Data 110, https://doi.org/10.1007/978-3-031-04903-3_7

1 Introduction

Currently, the flow of scientific information is growing every year, the main source of obtaining this information is information and communication technologies, which increase the intensity and quality of the learning process.

The development of information networks of the Internet gives us a wide range of opportunities for various services that allow us to combine the educational process in various training formats: e-mails, forums, Skype, chatting systems, wikis, blocks, etc. These forms are actively used and relevant nowadays, while there is the coronavirus infection, which has unexpectedly entered our life.

The modern development of society is based on a continuous learning process. Everyone has to study both at work and at the university, while the learning process becomes more and more complicated each time, a lot of information appears in various information networks. This implies the need for the development of new technologies in distance learning.

The traditional classroom system, with all its advantages, is limited in terms of meeting the growing needs of people. The modern person wants to have access to educational resources everywhere and at any time.

Modern digital multimedia and communication technologies have made it possible to overcome spatio-temporal boundaries in education. Distance learning technologies that were able to effectively adapt educational services to the modern rhythm of life and professional needs of people have arisen in response to these civilizational challenges.

Communication of teachers with students in educational institutions mainly takes place through the distance learning service (DLS) Moodle. This system makes it possible to create a single educational information space for students and teachers at the university.

The relevance of the research is that distance learning technologies allow using new forms of learning in educational activities, regardless of the location of the student, and the ability to constantly update the content of the educational process.

2 Methodology

In the process of researching distance learning technologies in the educational activities of the university, we set the goal to study the specificities of the use of distance learning techniques of the Moodle system on the basis of Minin University.

The following tasks were completed during the research process in order to achieve the set goal: to consider the main positive aspects of distance learning at the university, to analyze the effectiveness of using distance learning platforms.

With the development of scientific and technological progress, publications on distance learning appear in the scientific literature as a way of educational activity in educational institutions.

To solve the set research problem, we analyzed the publications of domestic scientists dealing with the issue of distance learning technologies—Samerkhanova [8], Bashmakov [1], Smirnova [9], Kondratenko [10] and other scientists [2, 3].

General method for exploratory data analysis and comparative analysis aimed at ensuring the effective use of all organizational, economic and regulatory resources of the use of information technologies in the educational activities of the university, served as a methodological framework of the research. Achieving this goal defines the effectiveness of the use of distance learning technologies at the university.

For practical implementation, a study aimed at the application of information technologies at Minin University was carried out. The process of teaching 2nd year students of the Faculty of Management and Social and Technical Services of Minin University using distance technologies in the Moodle system (https://edu.mininuniver.ru/course/index.php) [6] in the field of the training program 43.03.01 Service was considered as an object of research.

The study conducted a survey of 25 students on the advantages of distance learning in improving the quality of educational services and an analysis of the effectiveness of using distance learning technologies at a university.

3 Results

The use of modern tools for distance learning allows a change in the approach to the educational process itself, expanding the possibilities of interactive and problem classroom lessons. For example, the forms of lesson and extracurricular activities of students can be organized in a new way. Checking homework and completing tests can be transferred to an online format, and more attention can be given to creative teamwork in the classroom.

Remote technologies make it possible to intensify work with additional educational material and compensate for the lack of a library fund for educational and methodological literature.

In the process of studying the advantages of distance learning at the university, it can be highlighted that distance learning can be effectively used in the provision of educational services to homeschooled people or children with disabilities. And this is not a complete list of possible organizational and methodological benefits and advantages that modern distance learning technologies provide [5].

Today, many software products for organizing the educational process using distance technologies exist in the world.

Electronic courses on the Moodle platform were developed at the Faculty of Management and Social and Technical Services for teaching part-time students in the field of the training program 43.03.01 Service, on the basis of the requirements of the Federal State Educational Standard of Higher Education, in the correspondence course of study, most of the hours of contact work are devoted to the student's independent work.

The development of appropriate material on the content of disciplines with a certain amount of information, labor intensity and the use of distance technologies in the electronic educational environment is necessary for the implementation of the learning process of distance learning in a distance format.

Information and communication technologies are used, courses of disciplines «Service activity», «Organization and planning of activities of service enterprises», «Economic theory» and «Sales technology» are developed in the Moodle system for the implementation of these conditions at the university [3, 4, 7, 11].

Electronic courses have their own structure for its development based on the regulatory documents of the university.

The structure of the e-course includes standard elements of the educational process. University teachers develop lectures, practical tasks with examples of problem solving, topics of independent work, tasks for monitoring students' knowledge (tests, exams, test papers, etc.), based on the requirements for the content of electronic courses [6]. Each developed module of the discipline must include additional materials for obtaining information (presentations, hyperlinks, videos).

In the process of considering the technology of distance learning at the university, we have asked students about the benefits of distance education and improving the quality of educational services (Fig. 1).

According to the diagram, the main advantage is the use of additional resources in the Moodle system and this is 79%, the possibility of receiving advisory services of the teacher online is 67%, and the advantages of individual terms of mastering the discipline are 62%.

Thus, distance learning technologies have advantages in the possibility of using electronic resources.

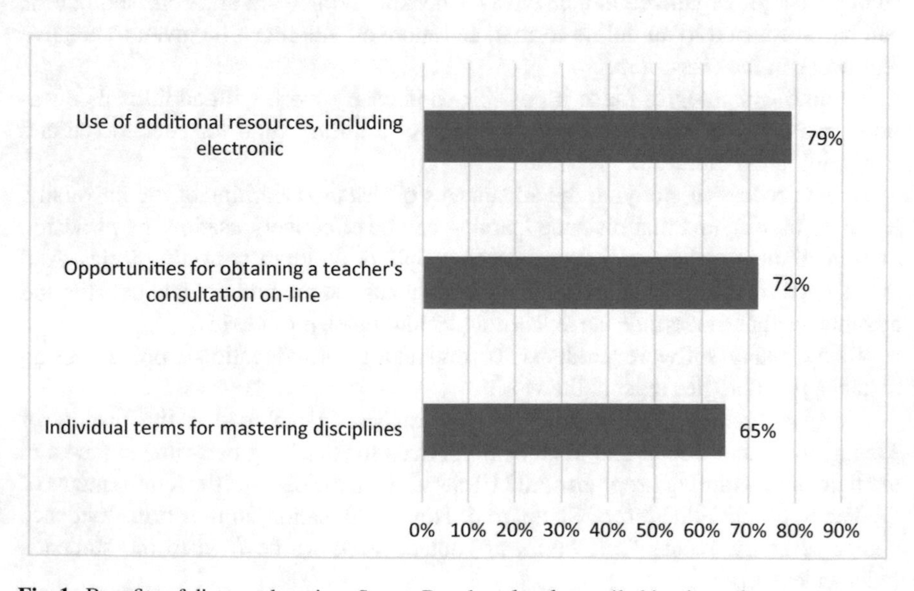

Fig. 1 Benefits of distance learning. *Source* Developed and compiled by the authors

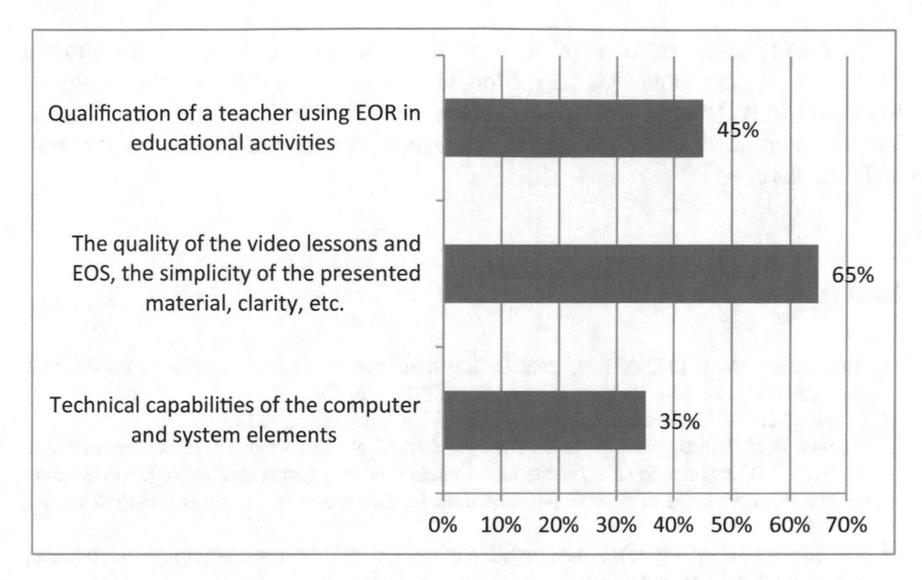

Fig. 2 Effectiveness of distance learning. *Source* Developed and compiled by the authors

According to students, the effectiveness of distance learning largely depends on the qualifications and level of training of the teacher, using the technical capabilities of the computer and elements of the Moodle system (Fig. 2).

According to students, 65% of the effectiveness of mastering the material with the use of e-learning is the quality of the content of the electronic educational resource, the ease of presentation of the material and its clarity.

When asked about the most effective tool for distance learning technology in e-courses, students noted the control of knowledge as 56%, the implementation of practical work as 28%, lectures (learning new material) as 16%.

After analyzing the data obtained, we can conclude that the efficiency of distance learning is quite high in general and the applied methods, ways, types, tools, directions and models of such training are also very effective.

4 Conclusion

E-learning is playing an accrescent role in the education system. First of all, the activation of this process is facilitated by the development of the Internet and web technologies, which provide new opportunities in the development of this form of studies.

The improvement of ICT in the field of education and the rapid decline in the cost of the services they provide create conditions when distance learning becomes not only affordable, but also a very attractive form of education.

This study analyzes the application of distance learning technologies through the electronic educational platform Moodle of Minin University at the Faculty of Management and Social and Technical Services. The results of the study showed the potency of the use of e-learning for students in the field of the training program 43.03.01 Service.

References

1. Bashmakov, A. I. (2005). Intelligent information technologies: Textbook for universities. Bauman Moscow State Technical University (MSTU), p. 302.
2. Vlados, M. (2016). Distance learning. Textbook for universities, p. 192.
3. Garina, E. P., Romanovskaya, E. V., Andryashina, N. S., Kuznetsov, V. P., & Shpilevskaya, E. V. (2020) Organizational and economic foundations of the management of the investment programs at the stage of their implementation. *Lecture Notes in Networks and Systems, 91*, 163–169.
4. Japarova Samal, M. (2020). Implementation of the Moodle distance learning system in higher education. LAP Lambert Academic Publishing, p. 112
5. Malushko, E. Y., Lizunkov, V. G. (2020). The e-education system as a tool to improve the competitiveness of a specialist in the digital economy. *Vestnik of Minin University, 8*(2). https://doi.org/10.26795/2307-1281-2020-8-2-3. Accessed 01 Dec 2021.
6. Moodle system. (2021). https://edu.mininuniver.ru/course/index.php. Accessed 01 Dec 2021.
7. Romanovskaya, E. V., Kuznetsov, V. P., Andryashina, N. S., Garina, E. P., & Garin, A. P. (2020). Development of the system of operational and production planning in the conditions of complex industrial production. *Lecture Notes in Networks and Systems, 87*, 572–583.
8. Samerkhanova, E. K., & Balakin, M. A. (2020). Preparation of managers of professional educational programs to work in the digital environment of the university. *Vestnik of Minin University, 8*(2). https://doi.org/10.26795/2307-1281-2020-8-2-4. Accessed 01 Dec 2021.
9. Smirnova, Z. V., Kuznetsova, E. A., Koldina, M. I., Dyudyakova, S. V., & Smirnov, A. B. (2020). Organization of an Inclusive Educational Environment in a Professional Educational Institution. *Lecture Notes in Networks and Systems, 73*, 1065–1072.
10. Vaganova, O. I., Chelnokova, E. A., Smirnova, Z. V., Mukhina, M. V., & Ponomareva, E. (2020). Organizing E-learning using. cloud technologies. *International Journal of Advanced Trends in Computer Science and Engineering, 9*(4), 4844–4848.
11. Yashin, S. N., Koshelev, E. V., Sukhanov, D. A., Kuznetsov, V. P., & Romanovskaya, E. V. (2019). Method selection of graphic-analytical justification of effective innovative projects in the industrial safety field Studies. *Computational Intelligence, 826*, 1097–1114.

State Regulation of the Economy by Industry Using Big Data in the GovTech

Improving the Application of Information Technology in the Economy of Service Organizations

Zhanna V. Smirnova, Olga T. Cherney, Zhanna V. Chaykina, Natalia S. Andryashina, and Valentina A. Sidyakova

Abstract The article discusses the application of information technologies in the economics of service organizations. The author analyzes the process of applying information technologies on the example of one service organizations. Enterprises in the service sector are more and more specializing in the use of computer technologies in their activities, despite the numerous problems of informatization of enterprises. This study examines the effectiveness of the use of automated computer systems in service organizations using the example of hotel service. Information technologies are considered in terms of information processes, technologies for obtaining information and information processing. In this study, an indicator of the economic efficiency of the use of information technologies is determined. On its basis, an analysis of indicators of financial results of hotels in Nizhny Novgorod was made. As part of our research, we have selected Hotel Nizhny Novgorod LLC, one of the hotels in Nizhny Novgorod. According to the management system, the main information systems used in service organizations have been identified in hospitality organizations. The analysis of financial indicators of profitability for 2019–2020 was carried out according to the calculations of the main indicators. We have highlighted the most famous software products of the computer system and automation of «Hotel Nizhny Novgorod» LLC in the process of researching the activities of this hotel in the framework of informatization of the control system. In the course of the research, it was concluded that information technologies optimize the organization's management system, increase the efficiency of employees in certain areas of employee activity.

Keywords Information technologies · Service activities · Economy · Efficiency · Profit

JEL Codes R11 · R12 · R 58 · Q13 · Q18

Z. V. Smirnova (✉) · O. T. Cherney · Z. V. Chaykina · N. S. Andryashina
Minin Nizhny Novgorod State Pedagogical University, Nizhny Novgorod, Russia
e-mail: z.v.smirnova@mininuniver.ru

V. A. Sidyakova
Institute of Food Technology and Design-branch of the Nizhny Novgorod State University of Engineering and Economics, Nizhny Novgorod, Russia

V. N. Ostrovskaya and A. V. Bogoviz (eds.), *Big Data in the GovTech System*,
Studies in Big Data 110, https://doi.org/10.1007/978-3-031-04903-3_8

1 Introduction

The current state of development of information technologies is actively being introduced into the life of society and into the production activities of service enterprises in our time. The introduction of information technologies into the activities of organizations determines the effectiveness of the economic profit of enterprises as a whole [1].

The implementation of the conditions for informatization of the management activities of enterprises requires solving a number of problems related to the training of employees of enterprises in computer literacy. The introduction of information technologies in the organization of service activities is caused by many problems arising in the process of informatization of an enterprise, such problems include financing for the process of training employees, payment for Internet resources, the number of users with access to the Internet, etc.

Enterprises in the service sector are more and more specializing in the use of computer technologies in their activities, despite the numerous problems of informatization of enterprises [2].

The rapid growth of the computer network increases its subscribers and the volume of information resources every year.

Service activity organizations such as: travel companies, consumer services organizations, hotel services, catering organizations, organizations of sports and recreation centers, educational organizations, etc. are switching to automation of work in their activities. Information programs help employees of enterprises to record the progress of documentation, to work on the transfer of documents at a distance, to coordinate orders, to create an electronic signature, to work with clients and many other functions [3].

This study examines the effectiveness of the use of automated computer systems in service organizations using the example of hotel service.

2 Methodology

To achieve this goal, it is necessary to analyze the methods of the concept of information technologies, the classification of information technologies, to analyze the use of information technologies in the organization of hotel service and to consider performance indicators, the use of information technologies of enterprises by type of activity [4].

Today, hotel service is one of the most profitable and rapidly developing areas of providing services to the population. There have been positive changes in the field of tourism and hotel services over the past year in Russia. Most of the overseas tours were closed due to coronavirus infection, and most of the country's population preferred to rest in Russian resorts, so this study made it necessary to improve the activities of hospitality service enterprises using modern information technologies.

The methodological basis of the study was the general methods of analysis and comparative analysis aimed at ensuring the effective use of all organizational, economic, technological and social resources of the organization of operational management in the provision of services to the population [5].

The organization receives positive results from improving information support: possible intangible benefits (quality of information, increased productivity, reduction of late payments, faster service).

Achievement of this goal allows satisfying the needs of the population in a high-quality, timely manner and in full. The study analyzed and assessed the hotel information flow management system for the practical implementation of the proposed approach of innovative solutions to the effectiveness of the use of information technologies.

3 Results

The use of information technologies in hospitality organizations is a strategic goal of the business, it is used to control the activity of the structure, finances, material flows and workplaces of employees. Within the framework of the functionality, such information technologies can be attributed to information management, which performs the range of management tasks of the organization [6].

The main task of management in hotel service organizations is the use of information technologies. The main tools of information technologies are: data transfer processes, search for information, information processing, information distribution. In this case the information systems of the organization and departments; divisions and partners of the organization for cooperation will be the object of solving problems.

According to the management system, it is possible to single out the main information systems used in organizations in the hospitality business: Hotel management packages (Russian hotel, Hotel-Olymp, Edelweiss, Barsum, Reconline, Meridian-1, Hotel-2000, KEI Hotel, Hotel-3), Hotel financial management packages (financial analysis, investment performance analysis) and general software products (Word, Excel, Power Point, Access, Other packages) [7].

In the modern hotel business, the areas related to information flows are one of the least studied, and their specificity is not clearly defined. The possibility of production activity will largely depend on the way of organizing management process and on the diagnostics of the system. The formation of economic potential is characterized by a number of indicators: finance, personnel, production and others. Our study is determined in improving the economic efficiency of the use of information technologies, thus it is necessary to analyze the indicators of the financial results of hotels in Nizhny Novgorod.

As part of our research, we have selected Hotel Nizhny Novgorod LLC, one of the hotels in Nizhny Novgorod [8].

You can analyze the financial indicators of profitability by year according to the calculations of the main indicators (Fig. 1).

A brief analysis of assets, changes in capital and reserves, as well as the sum of non-current and all assets of the organization are presented in Fig. 2.

Based on the results of the financial condition, we see that the organization pays great attention to work on the calculation of financial indicators and, of course, in such a situation, one cannot do without the help of information technologies and settlement programs that help to maintain financial control.

Great attention is paid to improving hotel management within the framework of optimization of organizational structure [9, 10].

In most cases, the reform of the management system and the modification of the applied technologies for working with the market are carried out in parallel, because they are interconnected and one is impossible without the other. The control system

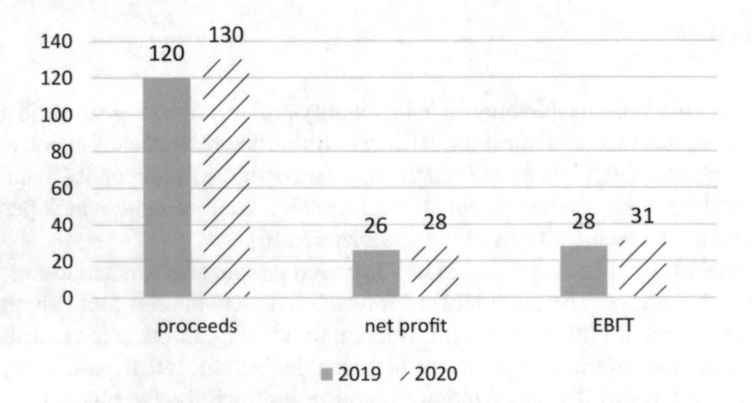

Fig. 1 Financial indicators, brief analysis. *Source* Developed and compiled by the authors

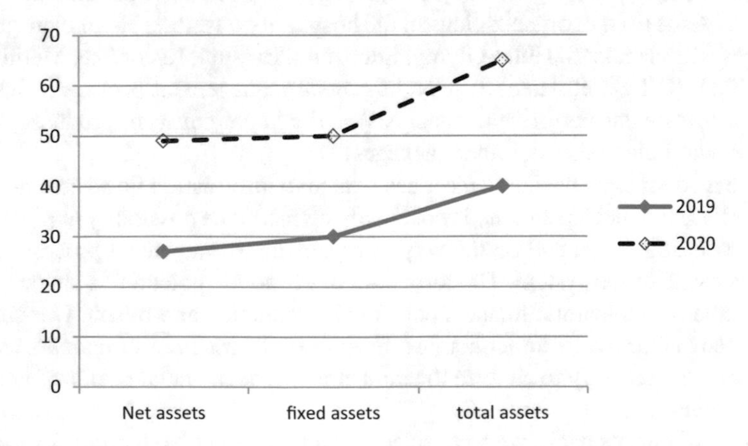

Fig. 2 Brief balance sheet analysis. *Source* Developed and compiled by the authors

is a problem area for optimization, because for this there is a significant limitation—the worldview of the top leader, which can be adjusted to a very insignificant extent [11].

Today, there is a rapid development of information technologies and software in the hotel business. Hotels and hotels need modern software and, of course, technologies to improve their business [12].

We have highlighted the most famous software products of the computer system and automation of «Hotel Nizhny Novgorod» LLC in the process of researching the activities of this hotel in the framework of informatization of the control system.

The Fidelio hotel automation computer system belongs to the most famous products of the German company Hrs, which is a manufacturer of automated systems for the hotel and restaurant business. Fidelio is a manufacturer of systems such as (Fidelio Front Office, Fidelio F & B Fidelio Food & Beverage) and Fidelio Eng. The Fidelio system is an integral part of the global computerized booking systems, such as Amadeus, Saber, Galileo, Worldspan, thus, all hotels presented in Fidelio are automatically uploaded to the global computerized booking system. The Fidelio computer system is one of the most popular systems for the hotel industry [13].

The Fidelio FO system makes:

- booking and registration of guests;
- accumulation of information about unpaid customer invoices coming from various points of sale;
- inclusion of information about non-cash payments;
- receiving financial and statistical reports.

The latest technologies Opera Enterprise Solution provides a unique opportunity to work in client–server application, as well as through Internet browser («thin client»). This specific line can significantly reduce costs at every stage of the hotel IT system lifecycle, including acquisition, installation, support and upgrades.

4 Conclusion

Thus, the use of information technologies in the business of service activities is very significant. After analyzing the role of information technologies in one of the organizations, we can conclude: information technologies are strategically important at the present stage of development of society, which means that these technologies will grow rapidly in the near future. Information technologies optimize the organization's management system, increase the efficiency of employees in certain areas of employee activity.

Informatization of the economy is the transformation of information into an economic resource of paramount importance. This happens on the basis of computerization and telecommunications, which provide fundamentally new opportunities for economic development, multiple growth in labor productivity, solving social and economic problems and establishing a new type of economic relations.

References

1. Belomoitsev, D. A. (2014). Basic methods of cryptographic data processing: tutorial. In D. A. Belomoitsev, T. M. Volosatov, & S. V. Rodionov. – M.: Bauman Moscow State Technical University, 80 p.
2. Belomoitsev, D. A. (2014). Basic methods of cryptographic data processing: tutorial. In D. A. Belomoitsev, T. M. Volosatov, S. V. Rodionov. - M.: Bauman Moscow State Technical University, 80 p.
3. Garina, E. P., Garin, A. P., Kuznetsov, V. P., Romanovskaya, E. V., & Andryashina, N. S. (2020). Definition of key competences of companies in the "product-production" system. *Lecture Notes in Networks and Systems, 73*, 737–746.
4. Garina, E. P., Kuznetsov, V. P., Egorova, A. O., Romanovskaya, E. V., Garin, A. P. (2017). Practice in the application of the production system tools at the enterprise during mastering of new products. In *Contributions to economics*, (№ 9783319606958, pp. 105–112).
5. Kulueva, C. R., Myrzaibraimova, I. R., Alimova, G. B., Kuznetsov, V. P., & Romanovskaya, E. V. (2020). The role of scientific and educational platform in formation of the innovative economy of Kyrgyzstan: Foreign experience, realities, and prospects. *Lecture Notes in Networks and Systems, 91*, 484–491.
6. Kuznetsov, V. P., Romanovskaya, E. V., Egorova, A. O., Andryashina, N. S., Kozlova, E. P. (2018). Approaches to developing a new product in the car building industry. *Advances in Intelligent Systems and Computing*, *622*, 494–501.
7. Lizunkov, V. G., Morozova, M. V., Zakharova, A. A., & Malushko, E. Y. (2021). On the issue of criteria for the effectiveness of interaction between educational organizations and enterprises of the real sector of the economy in the conditions of advanced development territories. *Vestnik of Minin University*, *9*(1).
8. Pankratov, F. G. (2015). Commercial activity: textbook for universities. In F. G. Pankratov (Ed.), 8th ed, rev. and add. - M.: Dashkov and Co. – 502 p.
9. Rudenko, I. V., Bystrova, N. V., Smirnova, Z. V., Vaganova, O. I., & Kutepov, M. M. (2021). Modern technologies in working with gifted students. *Propositos y Representaciones*, *9*. https://doi.org/10.20511/pyr2021.v9nSPE1.818. Published: JAN.
10. Saak, A. E. (2013). Information management technologies: T for universities. In A. E. Saak, E. V. Pakhomov, V. N. Tyushnyakov (Eds.), 2nd ed. M. St. Petersburg, 318 p.
11. Sedykh, E. P. (2019). Normative legal support system of project management in education. *Vestnik of Minin University, 7*(1)(26), 1.
12. Smirnova, Z. V., Vaganova, O. I., Gruzdeva, M. L., Golubeva, O. V., & Kutepov, M. M. (2020). Social and economic efficiency and quality of providing services to the population in the form of service activities. *Lecture Notes in Networks and Systems, 73*, 1029–1039.
13. Voronova, E. M., Lapshova, A. V., Bystrova, N. V., Smirnova, Z. V., Bulaeva, M. N. (2021). Organization of virtual interaction in the context of the coronavirus pandemic. *Propositos Y Representaciones*, *9*. Published: JAN. https://doi.org/10.20511/pyr2021.v9nSPE1.820.

Development of the Service Sector in a Digital Environment

Zhanna V. Smirnova, Marina L. Gruzdeva, Zhanna V. Chaykina, Elena V. Romanovskaya, and Olga T. Cherney

Abstract The article discusses the issue of using digital technologies in service organizations. The author studied the impact of digital technologies on the productivity of employees of service organizations. The aim of the research is to study the application of modern information technologies in the field of service activities on the example of enterprises in the city of Nizhny Novgorod. In the course of the research, theoretical approaches to the study of the problem have been explored and the development of the use of digital technologies in Russian and foreign service companies has been analyzed. The article shows the indicators of the use of information technologies in service organizations in Nizhny Novgorod. Criteria of the economic efficiency of various types of activities of enterprises were considered. According to the study of the data of the agencies of the Federal Statistics Service of Nizhny Novgorod, it was concluded that the use of information technologies by organizations by type of economic activity today has a number of problems at the federal level.

Keywords Service sector · Digital technologies · Technical and economic indicators · Economy

JEL Codes R11 · R12 · R 58 · Q13 · Q18

Z. V. Smirnova (✉) · M. L. Gruzdeva · Z. V. Chaykina · E. V. Romanovskaya · O. T. Cherney
Minin Nizhny Novgorod State Pedagogical University, Nizhny Novgorod, Russia
e-mail: z.v.smirnova@mininuniver.ru

M. L. Gruzdeva
e-mail: gruzdeva_ml@mininuniver.ru

E. V. Romanovskaya
e-mail: alenarom@list.ru

V. N. Ostrovskaya and A. V. Bogoviz (eds.), *Big Data in the GovTech System*,
Studies in Big Data 110, https://doi.org/10.1007/978-3-031-04903-3_9

1 Introduction

Today, the development of digital technologies is one of the most important issues in the development of the country's economy as a whole. The development of digital technologies in various sectors of the national economy is being introduced into the service development system every year. Information technologies reduce the work of transferring data of employees, reduce the cost of paper media, increase the relationship with consumers and government agencies using Internet technologies.

The capacities of the services of Internet resources allow interaction between employees at a distance, such technologies are used not only abroad, but have also become widely available in Russian service organizations [12].

The growth in the development of enterprises providing services to the population is observed at this stage of development of service activities in Russia. Information technologies are widely developing as technologies for digitalizing the economy as a whole at enterprises with the development of service organizations.

The relevance of this study is that the use of information computer technologies is the most important factor in increasing the efficiency of production processes of enterprises in the service sector today. The use of information resources in the activities of enterprises acquire a competitive advantage of survival in a crisis.

The purpose of the research is to study the application of modern information technologies in the field of service activities on the example of Nizhny Novgorod, to analyze and assess the efficiency of enterprises in the service sector, to assess the impact of digitalization on the sustainable development of enterprises in the service sector [3, 4].

2 Methodology

To achieve this goal, it is necessary to analyze theoretical approaches to the study of digitalization and the development of the use of digital technologies in Russian and foreign service companies.

The theoretical basis of the study is the materials of scientific research and regulatory documents of foreign and Russian scientists: Zhukova M.A. «Information technologies of management in service», Morozov M.A. and Morozov N.S. «Information technologies in SCS and tourism» [5].

The methodological basis of the research is the general methods of analysis and comparative analysis aimed at ensuring the effective use of all organizational, economic, and regulatory resources of the application of information technologies. Achievement of this goal makes it possible to determine the effectiveness of the use of information technologies in the organization of service activities. A study aimed at the application of information technologies at service enterprises in Nizhny Novgorod was carried out for the practical implementation of digitalization in the service sector.

Today, the economy is based on continuous enhancement, modernization and improvement of the new management system, introduction of new economic solutions in the field of information technology [7].

According to 2021, countries such as Denmark, Czech Republic, Germany, Austria and France remain the leaders in the development of digitalization in the service sector. In turn, Russia belongs to the countries of transition to a digital economy, and this hinders the development of many processes [6].

Thus, the introduction of information technologies gives impetus to the development of organizations in the service sector.

3 Results

Research on the application of information technology in service organizations in Nizhny Novgorod is based on the basic concepts of informatization and computerization. Currently, information technologies are rapidly developing in the field of creating electronic programs that contribute to the activation of various actions related to the transfer of information through databases, to the use of a program with calculation formulas and statistical indicators, to the use and confirmation of electronic signatures, which reduce the course of actions of paper documentation, and they also contribute to the implementation of the working document flow of organizations.

Informatization is scientific and technological progress aimed at creating optimal conditions for information needs and implementation of the organization's work processes.

The informatization process is a constant developing process that takes place in all areas of the economy and is closely related to the system of management and consumption of various goods and services.

In the course of the research, we carried out a statistical analysis of the significance of information and communication technologies in Nizhny Novgorod. Based on the data obtained, we can see a positive trend in an increase in organizations using information technology in their activities (Table 1).

Research data show that 98% of organizations with service detail are provided with computers. The most demanded directions are the directions of development of information networks 97%. The rate of organizations with their own website pages increased to 92% [2].

Organizations differing in economic activity and using information technologies in their development locate in Nizhny Novgorod. According to the Federal Statistics Service of Nizhny Novgorod it can be seen that the use of information technology by organizations by type of economic activity can be distinguished by the following results for 2021 (Fig. 1) [9].

According to the diagram, we can see that the greatest efficiency in economic indicators is the problem of limiting the Internet from lack of funds. One of the

Table 1 Indicators of the use of information technologies in service organizations in Nizhny Novgorod

	Indicator		
	Organizations using personal computers (%)	Organizations using e-mail, global information networks (%)	Organizations that had websites on the Internet (%)
2019	89	90	76
2020	91	94	85
2021	98	97	92

Source Developed and compiled by the authors

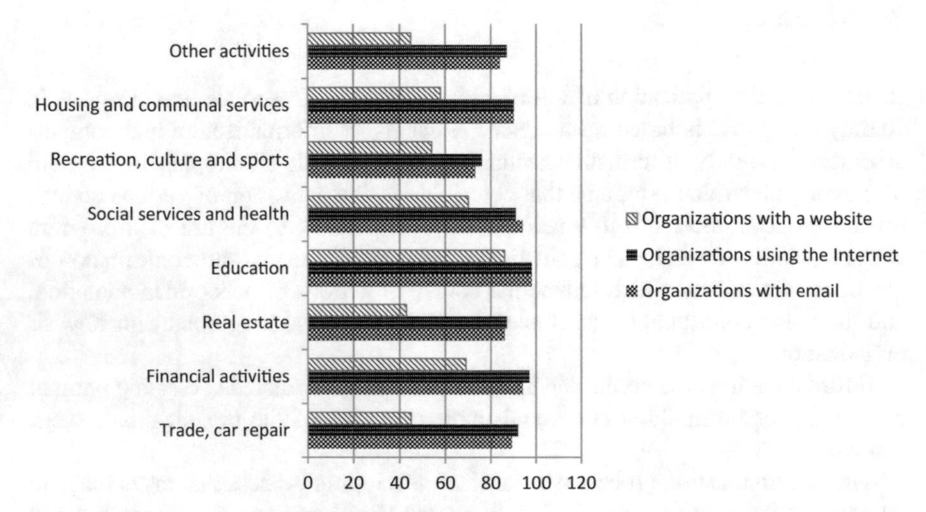

Fig. 1 Indicators of the use of information technologies in service organizations in Nizhny Novgorod. *Source* Developed and compiled by the authors

reasons for the low internet connection is the risk of fraudulent payment transactions. There is a lack of maintenance and work with Internet sources in organizations [10].

The development of informatization in organizations in the service sector is associated with an increase in the availability of working with information as a means of communication, and not for automating the business system.

Regular improvement of the management system in service organizations requires the development of internal communication network of employees and transfer of information on various types of enterprise activities. There is a need for the development of certain information programs for certain types of enterprise actions. The purchase of software control systems and qualified personnel who can work with these programs are required.

Today, the problems of professional training of workers in the framework of informatization of the activities of enterprises arise in organizations. The heads of enterprises practically do not allocate monetary funds for the retraining of their specialists in the field of digitalization of the activities of workers.

According to the statistical data of the study, the indicator of expenses for the informatization of enterprises' activities in the structure of expenses in the organizations of the service sector of Nizhny Novgorod increased almost 4,5 times.

In 2019 and 2021, costs only mainly for the purchase of computers and payment for communication services (50 and 42%) prevailed in the costs of digitalization of enterprises in various types of activities of organizations in Nizhny Novgorod. 4% in 2019 and 8% in 2021 was spent on training employees working on this equipment [1].

Based on the research, we see that the leaders of organizations do not want to train their employees in the framework of information technologies, and this does not suit the employer.

According to managers, cash costs for the purchase of computer equipment and special programs are necessary for the process of training employees in information technologies, which is not beneficial for the financial position of the organization [8].

Thus, the situation for the organization of new services in the field of maintenance and work with information systems is emerging. Despite all the financial difficulties, more than 700 thousand specialists work in the field of information technologies. About 2.3% of the country's population is employed in such industries in Russia, in comparison with the indicators of other countries, about 6%, it is low. Thus, there is a shortage of employees in the field of information technologies in the country [11].

4 Conclusion

Thus, the organizations of service activities in Nizhny Novgorod carry out a number of activities to introduce digital technologies into their production activities. Many organizations are actively involved in the design work of the development strategy of the Nizhny Novgorod region to create a digital space in organizations, taking into account the increasing needs and the possibility of gradual improvement of networks in connection with the emergence of new technical and software solutions.

References

1. Garina, E. P., Romanovskaya, E. V., Andryashina, N. S., Kuznetsov, V. P., & Shpilevskaya, E. V. (2020). Organizational and economic foundations of the management of the investment programs at the stage of their implementation. Lecture Notes in Networks and Systems, vol. 91, pp. 163–169.
2. Kolosova, O. Y. (2014). Socio-economic aspects of informatization of society. In Collections of Conferences SIC «Sociosphere». № 1. pp. 7–11.

3. Kuznetsov, V. Y., & Kuznetsova, E. I. (2012). Statistical study of informatization of economic activity as a factor of innovative activity. *Bulletin of the Orenburg State Agrarian University, 3*(35–1), 156–158.
4. Lizunkov, V. G., Morozova, M. V., Zakharova, A. A., & Malushko, E. Y. (2021). On the issue of criteria for the effectiveness of interaction between educational organizations and enterprises of the real sector of the economy in the conditions of advanced development territories. *Vestnik of Minin University, 9*(1).
5. Rezvushkin, S. V. (2014). Informatization of society and economic processes. *Annual Scientific Readings of the Branch of the Russian State Social University in Klin, 2*(12), 224–235.
6. Salyutina, T. Y., Kuzovkov, A. D. (2016). Analysis of methods and approaches to measuring the processes of informatization and movement towards the information society. *T-Comm: Telecommunications and Transport, 10*(6), 52–57.
7. Saykhanova, K. I. (2016). Informatization of society as one of the patterns of modern social progress. *International Journal of Humanities and Natural Sciences, 1*(1), 195–198.
8. Sedykh, E. P. (2019). System of normative legal support of project management in education. *Vestnik of Minin University, 7*(1), 26.
9. Smirnova, Z. V., Vaganova, O. I., Chanchina, A. V., Koldina, M. I., Kutepov, M. M. (2020). Development of research activity of future economists in the university. In Collection: Lecture Notes in Networks and Systems. Growth Poles of the Global Economy: Emergence, Changes and Future Perspectives. Plekhanov Russian University of Economics. Luxembourg, pp. 371–379.
10. Smirnova, Z. V., Kamenez, N. V., Vaganova, O. I., Kutepova, L. I., & Vezetiu, E. V. (2019). The experience of using the webinar in the preparation of engineering specialists. *Amazonia Investiga, 8*(18), 279–287.
11. Smirnova, Z. V., Kuznetsova, E. A., Koldina, M. I., Dyudyakova, S. V., Smirnov, A. B. (2020). Organization of an inclusive educational environment in a professional educational institution. Lecture Notes in Networks and Systems, vol. 73, pp. 1065–1072.
12. Stanchin, I. M. (2016). Information support of statistical research of enterprise finance. *Sustainable Development of Science and Education, 1,* 52–61.

Innovative Banking Services in the Conditions of Digitalization

Shakizada U. Niyazbekova, Gaukhar S. Kodasheva, Tamara Yu. Dzholdosheva, Makka G. Goigova, and Aigul A. Meldebekova

Abstract This work focuses on existing innovative products, online banking, Internet banking in commercial banks in Kazakhstan, France, Sweden, Brazil, and the UK. The rating of the cheapest banks in France in 2020 in terms of annual commissions for servicing via Internet banking is presented.

Keywords Banking innovations · Innovative products · Innovative technologies · Product line · Customer service quality

JEL Codes E43 · E59 · G21 · I20

1 Introduction

The actuality of this article lays in the fact that the bank of today's era is a competitive, stable bank with a convenient and attractive product line, which meets the needs of individuals and corporate clients. Modern banks pay special attention to optimizing existing processes, improving existing products and services, risk management systems, corporate governance issues, and also maintaining a strong position

S. U. Niyazbekova (✉)
Financial University Under the Government of the Russian Federation, Moscow, Russia
e-mail: shakizada.niyazbekova@gmail.com

Moscow Witte University, Moscow, Russia

G. S. Kodasheva
L.N. Gumilyov Eurasian National University, Astana, Kazakhstan

T. Yu. Dzholdosheva
M. Ryskulbekov Kyrgyz University of Economics, Bishkek, Kyrgyzstan

M. G. Goigova
Ingush State University, Magas, Ingushetiya, Russia

A. A. Meldebekova
Abai Kazakh National Pedagogical University, Almaty, Kazakhstan

V. N. Ostrovskaya and A. V. Bogoviz (eds.), *Big Data in the GovTech System*, Studies in Big Data 110, https://doi.org/10.1007/978-3-031-04903-3_10

in the post-crisis period. With the help of all of these factors, there is competition between commercial banks to retain and attract new ones.

Banks of foreign countries have already created basic innovations that have formed a modern technological order, based on microelectronics and informatics.

There has always been competition between banks, but the changing market for new technologies constantly imposes any new requirements on commercial banks, which are forced to master new types of devices, products, as well as expand financial transactions that meet the needs of the client, in accordance with an innovative policy that takes into account the level of banking risks for maximizing profits.

2 Materials and Methods

Such scientists as [3, 4, 14] studied consumer acceptance of online banking: expanding the technology adoption model, and [1, 2, 6, 8, 10–12, 15] considered the spread of online banking, product lines of innovative banking services.

The authors also used information from the website "Which of the following banking features would you be interested in using if your bank offered them?" in this article [17].

3 Results

As a result of the rather intensive growth of the domestic banking market, as well as the growth of competition, many commercial banks are forced to diversify their business, looking for free directions, without stopping the expansion of the list of services provided to clients. The active development of technologies for remote banking services contributes not only to the emergence of new additional banking services but also to the development of another area of strategic activity. Today's realities show that e-business is one of the most important trends and it can be one of the long-term. This will be critical for the continued development of modern technology for many years to come.

Experience of Kazakhstan

The era of the digital economy began with the opening of digital banks. Kaspi Bank JSC was one of them, which has a significant advantage over other banking systems by building an integrated financial electronic system for the provision of services [7, 9, 12].

Leading IT industry experts in Kazakhstan are considering the Kaspi.kz platform as the best digital service in the country. Banks in Kazakhstan have created many options for the basic value proposition of convenience, acceptability and security. To increase the number of payment cards, the country's commercial banks offer their customers additional loyalty programs, such as travel insurance, cash refunds,

Table 1 Awareness of customers about retail banking services and innovative financial solutions in commercial banks

No	Directions
1	Introducing clients to innovative financial solutions (Fintech)
2	Consumer awareness of financial technology
3	Financial technology consumer awareness by age
4	Consumer awareness of credit comparison sites
5	Consumer awareness of credit providers online only
6	Consumer awareness of crowdfunding
7	Crowdland consumer awareness
8	Consumer awareness of cryptocurrencies
9	Consumer awareness of alternative financial services
10	Consumer expectations of the future of Fintech in terms of age

Source compiled by the authors based on materials from [13]

discounts in stores, point's accumulation system, a free period of use, as well as gifts for the achieved turnover [3, 5].

Thus, Kazakhstani banks are modernizing payment cards and began issuing payment cards with a contactless function, as well as customers can pay for goods and services without entering a PIN code.

Table 1 presents the features of customer awareness of retail banking and innovative financial solutions.

Familiarity with mobile payment methods, an acquaintance of customers with banking and payment services, customer satisfaction with the bank, mobile payments, and others can be also included in this list.

Experience of France

According to the research, the cheapest bank in France in 2020 was Boursorama Banque because its service amounted to 26.41 EUR per annum for the year. Fortuneo took second place and ING took third place, these three institutions were virtual banks, completely switched to online banking services.

From 2008 to 2018, the proportion of the French population using internet banking services increased from 40% to more than 60%.

The practice of online banking and electronic banking system allows customers to make payments, as well as various transactions over the Internet and can be part of the services of traditional financial institutions and virtual banks. Among French people at age 25–34 years, almost 80% used online banking.

In addition to this, 40% of French people, who were surveyed in 2018, said that they had downloaded the bank's banking app to their mobile phone.

Digital payments and FinTech Online payment methods have proven to be quite popular with French consumers. According to a study conducted in France in early 2018, it turned out that they used online payment services, and about one-third of consumers used payment services between individuals. The Statista Market Outlook report on the FinTech market states that the digital payments segment in 2018 amounted to more than 71 billion EUR.

In July 2021, the Department of Statistics reflected in the report the cheapest banks in France in 2020 in terms of annual commissions for banking services in the area of Internet banking services [16].

Experience of Sweden

In Sweden, Bank ID was the most common among the 31–40 age groups for 2020. Bank ID is an electronic identification solution for verifying the identity of citizens and also for services, such as online payments.

Almost 99% of the population in this age group used Bank ID, and they were closely followed by respondents aged 21–30. It was least popular among Swedes aged 81 and older.

However, more than a third of them used this identification method. In terms of Bank ID account types, the mobile account was the most common in Sweden during the assessed period.

Types and sectors of services[1]

Internet and mobile banking totalled nearly 54; using Bank ID in Sweden in 2020. The second sector of commonly using Bank ID was the payment services sector, which accounted for approximately 17% of usage.

Digital payments have become popular in recent years. However, in 2019, card payment continued to be the most common payment method for online shoppers in Sweden. According to a 2019 survey, 18% of respondents used online banking.

Experience of the UK

Statista Research Data shows the proportion of smartphone owners using mobile banking apps in the United Kingdom (UK) in 2015, by different age groups.

In general, the use of mobile banking applications declines with the age of the user.

About 69% of smartphone owners aged 18–24 years have used mobile banking applications, compared with 30% of smartphone owners aged 50–64 years. The only age group not following this trend was 65 + , in which 35% of smartphone owners used mobile banking applications.

Experience of Brazil

According to the Statista Research Department, in 2021, the number of Brazilian Nubank in 2016–2027 had reached 40 million customers as of June 2021, up from the 1.3 million customers reported at the end of 2016. This digital bank was one of the first Brazilian unicorn companies and also the first Latin American startup, Decacorn.

[1] URL: https://www.statista.com/statistics/803158/interest-towards-novel-banking-features-usa-by-generation/ (Accessed: 06.10.2021).

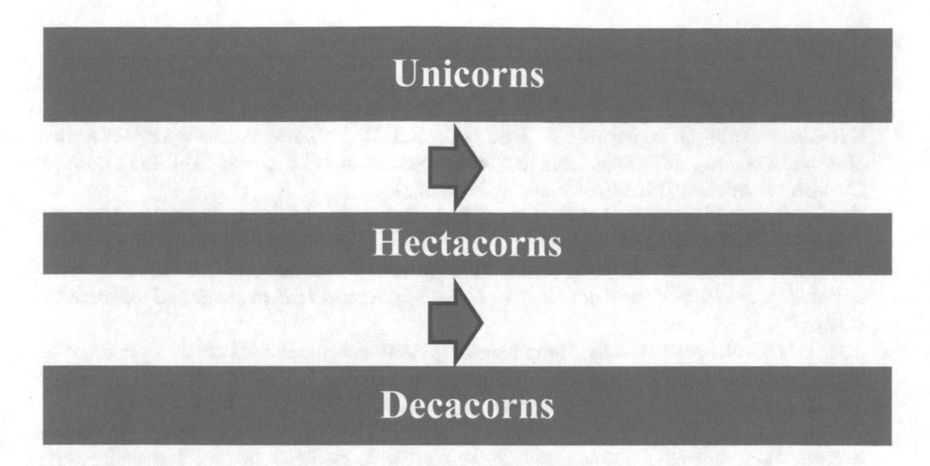

Fig. 1 Major cost categories of successful startups. *Source* Compiled by the authors

Categories of startups

Startups are defined as innovativc new businesses created by entrepreneurs seeking to influence and capture the market.

In terms of cost, successful startups can be divided into three main categories (Fig. 1).

Unicorns are private startups worth 1,000,000,000 USD or more.

Meanwhile, decacorns and hectacorns are private companies, the market value of which has reached more than 10 billion USD and 100 billion USD, respectively. As of April 2021, there were about 600 unicorn startups around the world, most of which are located in the United States and China.

4 Conclusion

Thus, many commercial banks of the countries are leaders in providing innovative services in the banking market; they have installed an iPhone application in the App Store, which provides information about the bank, in particular, news, exchange rates, addresses of branches and ATMs, as well as access to information about more than 100 shopping and service centers. Some banks offer cardholders up to 30% discounts on the purchase of various goods and services.

The main focus is on non-cash banking by providing remote banking services and simplifying various transactions for customers. The banks have been tasked with raising the level of awareness of the population in the area of long-distance communication services and non-cash payments.

References

1. Baidalinova, A. S., Baigireyeva, Z., & Myrkanova, A. (2021). Household food security in Kazakhstan. In E. G. Popkova & B. S. Sergi, B. S. (Eds.) *"Smart Technologies" for Society, State and Economy*, ISC 2020. Lecture Notes in Networks and Systems, (Vol. 155). Springer, Cham. https://doi.org/10.1007/978-3-030-59126-7_13.
2. Bradley, L., & Stewart, K. (2003). The diffusion of online banking. *Journal of Marketing Management, 19*(9–10), 1087–1109.
3. Brodunov, A. N., Ushakov, V. Y. (2015). Justification of financial decisions in conditions of uncertainty. *Bulletin of the Witte Moscow University. Series 1: Economics and management, 1*(12), 30–36.
4. Gavrilova, E. N. (2018). Credit history bureau: formation, evaluation of efficiency and ways of improvement. *Bulletin of the Moscow University named after S. Yu. Witte. Series 1: Economics and Management, 4*(27), 34–42. https://doi.org/10.21777/2587-554X-2018-4-34-42.
5. Gavrilova, E. N. (2019). Investment banking as a direction of banking activity: the essence, features and problems of development. *Bulletin of the S. Yu. Witte Moscow University. Series 1: Economics and Management, 4*(31), 81–86. https://doi.org/10.21777/2587-554X-2019-4-81-86.
6. Grekov, I. E., & Blokhina, T. K. (2016). The influence of macroeconomic factors to the dynamics of stock exchange in the Republic of Kazakhstan. *Economy of Region, 12*(4), 1263–1273. https://doi.org/10.17059/2016-4-26.
7. Ivanova, O. S., Suleimenova, B., Yerzhanova, S. K., & Berstembayeva, R. K. (2021). Oil and gas investment opportunities for companies in modern conditions. In E. G. Popkova, V. N. Ostrovskaya & A. V. Bogoviz (Eds.), *Socio–economic systems: Paradigms for the future.* Studies in Systems, Decision and Control, (Vol. 314). Springer, Cham. https://doi.org/10.1007/978-3-030-56433-9_70.
8. Kunanbaeva, K., Argyngazinov, A., Madiiarova, D., & Sagyndykova, R. (2020). Development of integration processes in the mining and metals sector in Russia and Kazakhstan under business transformation. In *E3S Web of Conferences*, (Vol. 210). https://doi.org/10.1051/e3sconf/202021013032.
9. Kurmankulova, R. Z., Anzorova, S. P., Goigova, M. G., & Yessymkhanova, Z. K. (2021). Digital transformation of government procurement on the level of state governance. In E. G. Popkova V. N. Ostrovskaya & A. V. Bogoviz (Eds.), *Socio–economic systems: Paradigms for the future.* Studies in Systems, Decision and Control, (Vol. 314). Springer, Cham. https://doi.org/10.1007/978-3-030-56433-9_69.
10. Maisigova, L. A., Isayeva, B. K., & Dzholdosheva, T. Y. (2021). Features of relations between government authorities, business, and civil society in the digital economy. In: E. G. Popkova, V. N. Ostrovskaya, A. V. Bogoviz (Eds.), *Socio–economic systems: Paradigms for the future.* Studies in Systems, Decision and Control, (Vol. 314). Springer, Cham. https://doi.org/10.1007/978-3-030-56433-9_144.
11. Moldashbayeva, L. P., Zhumatayeva, B. A., Mezentseva, T. M., & Shirshova, L. V. (2021). Digital economy development as an important factor for the country's economic growth. In E. G. Popkova, V. N. Ostrovskaya, A. V. Bogoviz (Eds.), *Socio–economic systems: Paradigms for the future.* Studies in Systems, Decision and Control, (Vol. 314). Springer, Cham. https://doi.org/10.1007/978-3-030-56433-9_38.
12. Nurpeisova, A. A., Smailova, L. K., Akimova, B. Z., & Borisova, E. V. (2021). Condition and prospects of innovative development of the economy in Kazakhstan. In E. G. Popkova, V. N. Ostrovskaya, A. V. Bogoviz (Eds.), *Socio–economic systems: Paradigms for the future.* Studies in Systems, Decision and Control, (Vol. 314). Springer, Cham. https://doi.org/10.1007/978-3-030-56433-9_184.
13. Official website Statista. Retrieved September 25, 2021 from https://www.statista.com.
14. Pikkarainen, T., Pikkarainen, K., Karjaluoto, H., & Pahnila, S. (2004). Consumer acceptance of online banking: an extension of the technology acceptance model.

15. Rakhimova, S., Kunanbayeva, K., Goncharenko, L., & Pigurin, A. (2019). Balanced system of indicators for the assessment of innovative construction projects efficiency. In *E3S Web of Conferences*, (Vol. 110, pp. 21–54). https://doi.org/10.1051/e3sconf/201911002154.
16. Rating of the cheapest banks in France in 2020 by annual commissions (EUR). Retrieved September 25, 2021 from https://www.statista.com/statistics/757523/cheapest-banks-annual-banking-fees-france/.
17. Which of the following banking features would you be interested in using if your bank offered them? Retrieved September 25, 2021 from https://www.statista.com/statistics/803158/interest-towards-novel-banking-features-usa-by-generation/.

Analysis of the FinTech Segment in the Russian Financial Services Market

Ekaterina A. Isaeva, Elena V. Fedotova, Inessa G. Nazarova, Evgeniya S. Tishchenko, and Louiza I.Iljina

Abstract The article aims at identifying the features of the innovative finance (Fintech) segment in the Russian financial services market, as well as analyzing its development and assessing changes under the influence of the pandemic. The authors reveal the structure of the FinTech segment and systematize the features of applied financial technologies in the sectors of payments and transfers, financing and lending, and capital management. The article highlights the competitive advantages of FinTech companies in comparison with the services of traditional financial and credit institutions. The authors evaluated scenarios for the development of the Russian financial services market (retention of control over the markets by traditional financial companies based on the accumulation of achievements of the innovative FinTech industry; division of the market into many narrow segments providing financial services while maintaining the positions of traditional major players; absorption by FinTech companies of traditional financial institutions and the development of their FinTech divisions). The article studies the changes in the development of the financial technology sector as a result of the impact of the pandemic. The authors proved that the efficiency of the functioning of innovative companies in the FinTech segment during the pandemic is higher than traditional ones. The factors hindering the development of the FinTech segment in Russian conditions are highlighted. The

E. A. Isaeva (✉)
Financial University Under the Government of the Russian Federation, Moscow, Russia
e-mail: economresearch@mail.ru; dipmesi@mail.ru

Moscow University for Industry and Finance "Synergy", Moscow, Russia

E. V. Fedotova
Kaluga Branch of the Russian State Agrarian University - Moscow Agricultural Academy Named After, K. A. Timiryazev, Kaluga, Russia

I. G. Nazarova
Ukhta State Technical University, Ukhta, Russia

E. S. Tishchenko
Kuban State Technological University, Krasnodar, Russia

L. I.Iljina
Pitirim Sorokin Syktyvkar State University, Syktyvkar, Russia

V. N. Ostrovskaya and A. V. Bogoviz (eds.), *Big Data in the GovTech System*, Studies in Big Data 110, https://doi.org/10.1007/978-3-031-04903-3_11

guidelines for improving the operating conditions of the Russian FinTech segment of the financial market are determined.

Keywords Financial services · FinTech · Innovative finance · Pandemic · Artificial intelligence

JEL Codes G14 · O33

1 Introduction

The current stage of development is associated with informatization and digitalization, which has affected all areas of people's lives [1, 2]. The financial sector is no exception. Technologies have the most diverse impact on financial services, and segments of the financial services sector are most susceptible to change [3]. The result of the penetration of innovations into financial markets was the formation of the FinTech segment, in other words, a set of companies implementing projects related to the introduction of digital technologies such as big data, artificial intelligence and machine learning, robotics, blockchain, cloud technologies, biometrics and others in the financial sector [4]. These are dynamic companies serving various sectors of the economy and using all kinds of innovative technologies in their activities [3]. The activities of companies in the FinTech segment contribute to the development of competition in the financial market, increase the availability, quality and range of financial services, reduce risks and costs in the financial sector, as well as increase the level of competitiveness of companies and technologies [5]. However, traditional companies operating in the financial services market have begun to experience threats from FinTech sector players due to the use of more advanced technologies by the latter, providing higher profits in value chains [6]. According to PwC forecasts, in the coming years, there will be a redistribution of the financial services market in favour of independent FinTech companies. Therefore, there is an urgent question of maintaining the normal activities of all participants in the financial services market to ensure its balance [3]. Another problem is maintaining the positive dynamics of the financial services market during the pandemic [7]. Currently, many powerful factors are exerting pressure on society: from epidemiological, economic and social changes to a shift in the balance of power in the global economy. However, the technological factor has a disproportionately strong impact on the financial services sector [3]. How modern innovative technologies will be able to ensure the sustainable development of the financial services market during the pandemic and support the well-being of all market participants, as well as preserve the quality of services provided to consumers?

2 Literature Review

The essence of innovative financial technologies, as well as models of their implementation, are described in the works of [6, 8–15]. The authors emphasize that the use of innovative technologies is accompanied by organizational and product innovations. New technologies are being actively introduced both by traditional financial market players (banks, investment, insurance and other companies) and by financial and technological structures specially created for their use. Belozyorov et al. [8] determine that the stability of traditional financial companies under the pressure of new players is determined by the degree of influence of various factors, for example, consumer inertia, overregulation of the sector, complex market mechanisms. In addition, traditional financial companies can develop effective competitive strategies to maintain their position in the market. According to the prevalence of certain factors, various scenarios for the development of financial markets can be implemented [16, 17]. The impact of the pandemic on the socio-economic reality attracts the attention of researchers. Despite the relative novelty of the problem, there are already a lot of studies in this area in 2020. In particular, experts point out that FinTech has become one of the beneficiaries of restrictive measures during the fight against the pandemic. The acceleration of the transfer of business operations and trade to the online environment and the growth in the number of users of remote services have led to a race of capitalization of digital financial services [18]. Ulybina and Bogatyreva [14] note an increase in the population's demand for cashless payments during the pandemic.

3 Methodology

The purpose of the study is to identify the features of the innovative finance (fintech) segment in the financial services market, to analyze its development in Russian conditions, as well as to assess the ongoing changes under the influence of the pandemic.

Research objectives: (1) to reveal the structure of the FinTech segment and the features of the applied financial technologies; (2) to analyze scenarios for the development of the Russian financial services market; (3) to investigate changes in the development of the financial technology sector as a result of the impact of the pandemic.

Research methods: method of theoretical analysis, graphical method, method of economic analysis, induction and deduction, systematic approach.

4 Results

Let's consider the structure of the FinTech segment and highlight the features of the applied financial technologies: (1) Payments and transfers: online payment services, online transfer services, P2P2 currency exchange, B2B3 payment and transfer services, cloud cash desks and smart terminals, mass payment services; (2) Financing and lending: P2P consumer lending, P2P business lending, crowd-funding; (3) Capital management: robo-advising, financial planning programs and applications, social trading, algorithmic exchange trading, target savings services and other [19].

A third of the entire FinTech market accounts for payments and transfers. This is because, in the conditions of active development of digital interaction between individuals, the growth of e-commerce (the volume of the Russian e-commerce market reached 30.6 billion dollars in 2019 or 1.3% of GDP) and digitalization of traditional organizations convenient instant cashless payments prove their obvious advantages in the financial market. The degree of digitalization of key participants in the money transfer market in Russia indicates their willingness to compete with banks and technological services, experiment with other channels and tools and meet the growing influence of regulatory authorities [20]. Settlements between financial organizations, business organizations and Internet users in the process of buying and selling goods and services via the Internet can be carried out both directly through remote financial services (mobile payment services (mobile banking, SMS banking, mobile operator payments, NFC payments); non-mobile banking services (Internet banking); non-bank non-mobile services (electronic money systems)), and through a payment aggregator. Online money transfers in Russia are carried out by various financial organizations. For example, Sberbank of Russia and Zolotaya Korona have the most balanced proportion of online presence with a noticeable gap from other market participants. Transfer from card to card online is offered by Blizko, Contact, Western Union, Unistream, Sberbank of Russia. Western Union, Golden Crown, Unistream have mobile applications. The most active position in social networks is occupied by Contact, Western Union, Golden Crown, Unistream. The payment aggregator allows sellers to connect several payment methods on their website at once: bank cards, virtual money, cash when paying for an order to a courier or at a pick-up point, payment for purchases from a mobile phone account, as well as by bank transfer, etc. Examples of payment aggregators are Yandex.Cash Register, WebMoney Transfer, PayMaster, RoboKassa, Qiwi and others. Bank cards remain the most popular means for payments on the Internet. In 2019, 90.5% of Russians used them. 89.7% paid via Internet banking, 77.6% paid with electronic money. The volume of transfers between individuals in 2019 amounted to 38.3 trillion rubles, which is 17.4% more than in 2018. People transfer money between their accounts—50.9% of the annual volume of P2P transfers, the rest is regular and irregular transfers to loved ones, joint payment of bills, and so on. 38% of the volume of transfers to other people is payment for goods and services [21].

The second part of the FinTech segment consists of companies operating in the field of financing and lending. The intensive development of such companies began in the crisis of 2008 when the requirements for banks increased, and numerous companies lost the opportunity to lend to their projects. FinTech segment companies are engaged in financing high-risk projects, small and medium-sized businesses, individuals[4]. This also includes P2P lending, where lenders are not financial organizations, but users of FinTech services. In addition, FinTech companies allow people with interesting ideas to quickly and easily attract financing. In particular, crowdfunding allows participants to raise the necessary funds by attracting interested investors [22]. There are 26 official crowdfunding platforms in Russia. The most famous of them are Planeta.ru, Boomstarter, Stream, City of Money. In 2019, the total volume of transactions using such platforms amounted to more than 97.9 million dollars, while more than 60% of the funds were raised through crowdfunding.

The next group of companies in the FinTech segment includes companies engaged in capital management. Their activities are related to robo-advising (financial advice and service for creating and managing an investment portfolio with minimal human intervention), the creation of financial planning programs and applications, social trading (a form of network interaction that unites traders, investors and analysts into a single interactive environment), algorithmic exchange trading, target savings services, virtual insurance, infrastructure development of the FinTech segment (ensuring the security of customer data, uninterrupted operation of payment services, working with big data, as well as providing information to clients about possible services of the FinTech industry). The Russian market of robo-advisers has a restrained dynamics of development. Some Russian services are not sufficiently transparent and accurate in the process of determining the level of risk tolerance. However, large robo-advisors are associated with banks or management companies, offer their investment products or partner products as assets. Such services include: Financial Autopilot (Finex), personal financial assistant (Alfa-capital), simple investments (Sberbank + Finex), VTB Autopilot (VTB + Finex). Nevertheless, the portfolios of such robo-advisors are noticeably losing in terms of risk/return ratio, diversification, mutual correlation of securities, the size of commissions, etc. In many industries, the share of algorithmic trading is increasing due to the demand on various platforms for the speed of work. Due to the development of new technologies and the growing number of HFT companies using high-frequency trading, the number of such transactions from the total trading volume on the FORTS futures market of the Moscow Exchange has changed significantly. Moreover, due to the advent of an automated trading system (ATS), financial markets are increasingly becoming dependent on the activities of hyperactive trading robots [23].

According to the results of 2019, the share of Russian insurance companies that use IT solutions related to the use of the Internet in the process of selling insurance products is close to 100%. Only 6% of the total numbers do not use Internet solutions when concluding insurance contracts. The leading segment is car insurance, it has dropped in terms of the share of electronic insurance premiums to the lowest value in history—24.3% for 2019. The largest types of insurance by the share of premiums for 2019 were also insurances of property of individuals (20.2%) and insurance of those travelling abroad (16.7%). A significant increase in the share of contributions for 2019 is noted for life insurance, mortgage and other types of credit insurance, as well as financial risk insurance. The total volume of the electronic insurance segment, excluding compulsory motor liability insurance, reached 17 billion rubles in 2019. Social trading in Russia is not very developed. So, in 2007, FINAM launched the project comon.ru, it remains the largest transaction repetition service in Russia both in terms of the number of connected Russian customers (more than 5 thousand people) and in terms of funds (about 9 billion rubles) [24].

It is obvious that the services of companies in the FinTech segment have competitive advantages compared to the services of traditional financial and credit institutions: the ability to connect from any device anywhere; increasing technological capabilities for the most convenient, understandable and complete receipt of additional required data; the capacity to attract promising qualified specialists to perform tasks and provide competitive advantages; the ability to model, predict and analyze any direction of industry development; constant updates of both data and model versions; cost reduction by simplifying old systems, etc. [14]. Experts are considering several scenarios for the development of financial markets in Russia [8]. According to the first scenario, traditional financial companies (banks, insurance firms and other intermediaries) will be able to maintain control over the markets, accumulating the best achievements of the innovative FinTech industry. In this case, it is most likely that financial companies will implement an augmented conservative model that will improve the quality of interaction with consumers of traditional financial services, as well as the reliability of internal business processes (for example, by introducing a document management system based on blockchain, etc.), and an ecosystem of FinTech projects will form around the financial institution that complements the services of traditional financial organizations. This combination is currently being demonstrated by Sberbank and Alfa-Bank. The second scenario assumes that the market will be divided into many narrow segments providing financial services and focusing on private social, psychological, economic and geographical needs of consumers while maintaining the positions of traditional major players. Traditional financial institutions are gradually absorbing FinTech companies and creating their FinTech divisions, while "pure" FinTech operators are shifting towards an augmented innovation model by acquiring elements of traditional financial infrastructure (as the example of Tinkoff Bank shows). Indeed, organizations strive to maximize the benefits of combining the positive aspects of innovative and traditional models, while levelling the disadvantages of each of them. The third scenario suggests that digital multinational corporations will actively grow and displace traditional financial market players. This correlates with the factors of the evolution of Fintech: qualitative and quantitative growth of information technology; transformation of consumer habits in terms of widespread use of the Internet; improving the efficiency of financial organizations (seeking to reduce their costs through the introduction of remote service and automation, improve the quality of services and adapt the service delivery model to the changed needs of consumers) and non-financial companies (providing a higher level of monetization of their target audience by providing them with an increasing number of online services) [6, 25].

The pandemic is a factor determining which scenario will be implemented in Russian reality.

(1) There has been an increase in activity in the payments and transfers sector. In 2020, it became clear that FinTech technologies can be a tool for additional monetization of the e-commerce market. In the conditions of self-isolation and closure of physical stores, the Internet turned out to be the main channel for Russians to buy many categories of goods. Many companies have been forced to step up activities regarding the transfer of their business online, or significantly accelerate its digitalization. It has formed stable patterns for future development—the development of marketplaces, the explosive growth of the ready-to-eat and grocery delivery segment, the transformation of social networks and messengers into a full-cycle sales channel (from advertising to payment and delivery), integration into online sales of the self-employed, the development of a model for the sale of goods and services by subscription. The volume of online trade, in 2020, increased by 26%, from 2.9 trillion rubles to 3.7 trillion rubles. The development of e-commerce was ensured by increased demand in the commodity segment of the market, the volume of which increased by 55% from 1.9 trillion rubles to 2.9 trillion rubles at the end of the year [26]. In terms of the pandemic, banks expanded the offer of payment services to the population and business entities, paying special attention to the development of digital technologies. Non-cash transactions in 2020 amounted to 56.0 billion of 914.2 trillion rubles, an increase of 20.0% in quantity and 9.2% in volume compared to the previous year. Transactions of individuals grew at a faster pace (by 21.1% in number and by 20.1% in volume), which indicates the active use of non-cash payment instruments and services by the population[27]. The pandemic has given additional acceleration to the development of remote channels and contactless payments. In conditions of self-isolation and quarantine restrictions, accounts with remote access have become even more popular for customers of credit institutions. The rate of their growth (11.0%) almost doubled compared to the previous year—their number amounted to 290.0 million accounts as of 01.01.2021, and the share of active accounts of credit institutions' customers was almost 90%. The number of transactions made by customers of credit institutions using electronic technologies increased by 21.1% in 2020 (to 54.8 billion orders), and the volume—by 9.1% (to 818.2 trillion rubles) (Fig. 1).

Legal entities that are not credit institutions have made 2.3 billion transfers using electronic technologies of 729.1 trillion rubles (an increase of 3.8 and 7.5%, respectively). Payment orders sent via the network "Internet", accounted for 63%. Payment cards have become for most Russians a familiar payment instrument for everyday payment of goods and services, making transfers. Contactless technologies continued to develop actively. In 2020, almost 70% of payment cards issued by credit institutions supported the contactless payment function. On average, there were 1.4 contactless cards per inhabitant (at the beginning of 2020–1.2) (Fig. 2).

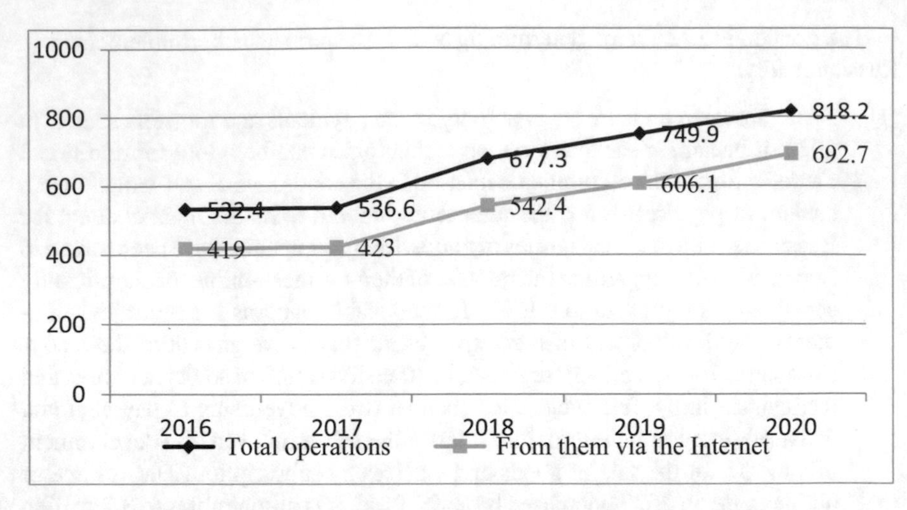

Fig. 1 The volume of transactions made by customers of credit institutions of Russia using electronic technologies, 2016–2020. *Source* Compiled by the authors based on [27]

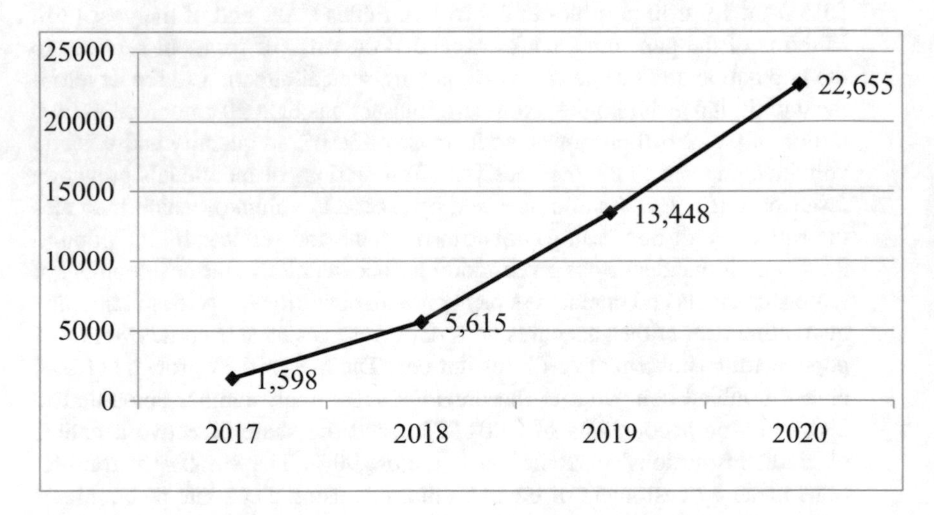

Fig. 2 The volume of transactions using contactless technologies in Russia, billion rubles, 2017–2020. *Source* Compiled by the authors based on [27]

The development of the regional payment infrastructure has served to spread non-cash payments throughout the country. In 2020, the positive dynamics in the number of POS terminals remained in all regions of the Russian Federation. At the same time, in almost a third of them, the growth rate of such devices amounted to 25% or more. In 2020, in 29 regions (against 9 regions in 2019), their number per 1000 inhabitants exceeded 25 devices.

(2) Online lending, P2P lending, crowdfunding allow increasing the coverage of the population with financial services by issuing loans to people with insufficient credit history and providing guarantees using new data sources (applications of client smartphones, based on the history of online sales and purchases) [8]. This became especially relevant during the pandemic when the material well-being of various categories of the population and economic entities was shaken. The COVID-19 pandemic has left an imprint on the business models of microfinance organizations and, against the background of the lockdown, accelerated the transition of the business to an online format. The portfolio of online microloans for the second quarter of 2020 decreased by only 1%, which was rather due to the high uncertainty of the situation. The preventive tightening of scoring parameters among online players and the subsequent intensification of competition in the spring—summer of 2020 provoked a price war for quality customers in online channels. In September 2020, despite the weakening of the isolation regime, the remote demand for loans secured by collateral remained at an increased level compared to the beginning of the year. The shares of online applications for car loans increased the most (from 21 to 30%) and mortgages (from 21 to 27%) [28]. In general, the pandemic has reduced the profitability of the main players in the Russian FinTech segment. Nevertheless, the efficiency of using assets and capital of digital companies is higher than that of traditional ones. The leaders in profitability are still Tinkoff Bank and UMoney. However, in 2020 they were overtaken by Alfa-Bank. In addition to the crisis, the algorithms of monetization of banking business are also influenced by the FinTech initiatives of the Central Bank. For example, the income of banks from transfers from card to card, as well as proceeds from acquiring, may be reduced due to the ability to pay using QR codes through a quick payment system. In 2020, the average size of the operation fell by about 1.5 thousand rubles and amounted to 7.3 thousand rubles. The number of transfers of individuals through the system of fast payments has increased by 12 times, and their volume—by 10 times [29]. The total volume of transactions concluded using crowdfunding platforms in Russia in 2020 amounted to more than 7 billion rubles, approximately the same it was in 2019. Despite the preservation of volumes in the pandemic 2020, the low popularity of investment platforms in Russia is explained by the fact that the investment efficiency remains lower than expected, and the mechanism itself carries high risks, including those associated with a low level of competence of retail investors.

(3) The pandemic has given impulse to the growth of the volume of operations and the customer base of FinTech projects, which are currently competing with both traditional banks and large technology companies and trading ecosystems developing their financial instruments. Due to the measures taken by the authorities in the fight against coronavirus, residents of the country have become 72% more likely to use FinTech applications not only for money transfers but also for money management. The global industry of automated robotic investment advisors (robo-advisers) in 2020 reached $1.4 trillion, an increase of 47% since 2019. Currently, most of the deals are in the US ($1 trillion),

China ($312 billion), the UK ($24 billion), Germany ($13 billion) and Canada ($8 billion). In Russia, the number of individuals as participants in the stock market is growing rapidly: in March 2020, investors bought Russian shares listed on the stock exchange of 43.7 billion rubles. The most massive inflow was recorded in the week from March 2 to 6, 2020 in the total amount of 30.4 billion rubles. The total inflow increased by 149.4 billion rubles over the past 12 months. Large robo-advisors are associated with management companies or banks and offer their investment products or partner products as assets[30]. Those robo-advisers that are available in mobile applications are becoming more and more successful, for example, Tinkoff Investments, Alfa Investments, Sberbank Investor, Yammi [31]. Global trends in the development of social trading have not spared the Russian financial market, although its results are modest. To date, the volume of issuance of such bonds has reached 125 billion rubles, 100 billion rubles accounted for the securities of Russian Railways. Shares of five ESG mutual funds with assets of about 8 billion rubles are traded on the Russian market. In particular, Tinkoff Investments launched the Signal social trading platform in synergy with Pulse, the largest social network for investors. The service allows you to follow strategies created based on the best investment ideas from market professionals [32].

5 Conclusions

Analysis of the dynamics of the development of the FinTech segment of the financial market during the pandemic shows that the efficiency of digital companies is higher than traditional ones. However, the development of FinTech in Russia is still hampered by the lack of the necessary regulatory framework for bringing some innovations to the market. To a certain extent, the development of the Russian FinTech segment is hindered, for example, by the closure of universities, their inability to bring promising developments to the end consumer. There are unresolved issues in the market of robo-advisers, related to the lack of the possibility of obtaining qualified automated advice on the selection of an investment portfolio without transferring funds to management, the lack of significant portfolio diversification, including both foreign and Russian assets, the lack of the possibility of choosing a currency for the investor to sell his accumulated and multiplied funds at the stage of information collection. The growing volume of information and its asymmetry complicate investment decision-making, increase transaction costs and affect the effectiveness of investment decisions made by households. This increases the need to develop formal methods that allow analyzing the situation, carrying out a criteria-based, scientific selection of financial instruments, rationalizing the choice made by investors, optimizing investments in financial instruments, as well as effective portfolio management. It is necessary to increase the financial literacy of market participants to expand the range of investment decisions made. Currently, the development of the architecture of systems for connection from any device anywhere and maximum attention to cybersecurity are particularly relevant.

References

1. Karpunina, E., Kharchenko, E., Mikhailov, A., Nedorezova, E., & Khorev, A. (2019). From digital development of economy to Society 5.0: Why we should remember about security risks? In *Proceedings of the 34rd IBIMA Conference* (pp. 3678–3688). Madrid, Spain.
2. Molchan, A., Karpunina, E., Kochyan, G., Petrov, I., & Velikanova, L. (2019). Effects of digitalization: New challenges for economic security systems. In *Proceedings of the 34th IBIMA Conference* (pp. 6631–6639). 13–14 November 2019. Madrid, Spain.
3. PwC. (2020). Financial services technologies in 2020 and beyond: Revolutionary changes. https://www.pwc.ru/ru/banking/publications/_FinTech2020_Rus.pdf. Accessed 10 Nov 2021.
4. Nikitina, T., Nikitin, M., & Galper, M. (2017). The role of FinTech segment companies and their place in the financial market of Russia. *Bulletin of St. Petersburg State University of Economics, 1–2*(103), 45–48.
5. Bank of Russia. (2021). Development of financial technologies. https://cbr.ru/fintech/. Accessed 10 Nov 2021.
6. Kotlyarov, I. (2018). Fintech: Essence and models of implementation. *ECO, 12*, 23–39.
7. Karpunina, E., Butova, L., Sobolevskaya, T., Badokina, E., & Pliusnina, O. (2021). The impact of the Covid-19 pandemic on the development of Russian national economy sectors: analysis of dynamics and search for stabilization measures. In *Proceeding of the 37th IBIMA Conference* (Vol. 1–2, pp. 1213–1226). Cordoba, Spain.
8. Belozyorov, S., Sokolovska, O., & Kim, Y. (2020). FinTech as a precondition of transformations in global financial markets. *Foresight and STI Governance, 14*(2), 23–35.
9. Kudryavtseva, Y. (2017). Banking services market: From the present to the future. *Financial Analytics: Problems and Solutions, 10*(4), 435–448.
10. Kuznetsov, V. (2017). Crowdfunding: Current regulatory issues. *Money and Credit, 1*, 65–73.
11. Nicoletti, B. (2017). The future of FinTech. *Integrating Finance and Technology in Financial Services*. London, Palgrave Macmillan.
12. Sigova, M., & Khon, O. (2017). Digital banking in Russia: The mainstream of FinTech. *Academic Notes of the International Banking Institute, 2*, 44–55.
13. Trofimov, D. (2018). Financial technologies in the field of retail payments: Development trends and prospects in the EU and Russia. *Economic Issues, 3*, 48–63.
14. Ulybina, L., & Bogatyreva, A. (2021). Financial technologies in a pandemic. *International Journal of Humanities and Natural Sciences, 6–1*(57), 234–238.
15. Yi, H. (2017). *SME Financing & FinTech in Korea*. Korea, Korea Technology Finance Corporation.
16. Burton, N., Bach, N., Calvin, P., Meissner, D., & Sarpong, D. (2019). Understanding cross border innovation activities: The linkages between innovation modes, product architecture and firm boundaries. *Journal of Business Research*. https://doi.org/10.1016/j.jbusres.2019.05.025. Accessed 10 Nov 2021.
17. Cambridge Center for Alternative Finance. (2018). Future of finance is Emerging: New Hubs, New Landscapes. Global FinTech Hub Report. Cambridge (UK): University of Cambridge.
18. Sberpro. (2021). Finance 2.0: How FinTech startups have grown into a Pandemic. https://sber.pro/publication/finansy-2-0-kak-fintekh-startapy-vyrosli-v-pandemiiu. Accessed 10 Nov 2021.
19. Bank of Russia. (2018). Main directions of financial technology development for the period 2018–2020. https://cbr.ru/Content/Document/File/84852/ON_FinTex_2017.pdf. Accessed 10 Nov 2021.

20. Tripadvisor. (2021). Money transfers (Russian market). https://www.tadviser.ru/index.php/Статья:Денежные_переводы_%28рынок_России%29. Accessed 10 Nov 2021.
21. Frank, R. G. (2019). Frank RG has studied the market of p2p transfers. https://frankrg.com/9088. Accessed 10 Nov 2021.
22. Rbc. (2021). What is crowdfunding: Platform overview and tips for beginners. https://trends.rbc.ru/trends/innovation/60a4f17d9a79473292bfd627. Accessed 10 Nov 2021.
23. Timofeev, A., & Lebedinskaya, O. (2017). The market is preparing for algorithmic trading. *Transport Business in Russia, 5*, 57–59.
24. Habr. (2016). Social trading: Follow the strong. https://habr.com/ru/article/300232/. Accessed 10 Nov 2021.
25. Filimonova, N., Ozerova, M., & Ermakova, I. (2017). Development of crowdfunding in agriculture of Russia. *Agro-Industrial Complex: Economics, Management, 7*, 68–77.
26. Rbc. (2021). Rating of payment services in Russia. https://marketing.rbc.ru/articles/12647/. Accessed 10 Nov 2021.
27. CBR. (2020). Results of monitoring in the national payment system for 2020. https://cbr.ru/content/document/file/124727/results_2020.pdf. Accessed 10 Nov 2021.
28. Comnews. (2020). Lending has gone online. https://www.comnews.ru/content/209548/2020-10-14/2020-w42/kreditovanie-ushlo-onlayn. Accessed 10 Nov 2021.
29. Bloomchain. (2021). Banking FinTech in Russia: Gaining altitude. https://bloomchain.ru/detailed/bankovskii-finteh-v-rossii-nabiraem-vysotu. Accessed 10 Nov 2021.
30. Garifova, L., Vakhitova, T., & Zulfakarova, L. (2020). Financial and credit system, budgetary, currency and credit regulation of the economy, investment resources. *Problems of Modern Economy, 2*(74), 177–180.
31. Deloitte. (2020). The expansion of RoboAdvisory in Wealth Management. Robo-Advisor 4.0. https://www2.deloitte.com/ru/ru.html. Accessed 10 Nov 2021.
32. Vedomosti. (2021). Tinkoff Investments launched the social trading platform “Signal”. https://www.vedomosti.ru/press_releases/2021/11/02/tinkoff-investitsii-zapustili-platformu-sotsialnogo-treidinga-signal. Accessed 10 Nov 2021.

Spatial and Territorial Factors in the Development of Communal Infrastructure Systems

Svetlana B. Globa, Dmitry V. Zyablikov, and Viktoria V. Berezovaya

Abstract The purpose of the work is to study the spatial and territorial aspects that affect the development of communal infrastructure, as well as to ensure an uninterrupted and high-quality supply of communal resources and a comfortable living environment for the population of the territory. It is necessary to give an objective and balanced assessment of the impact of these aspects in order to form an organizational and economic mechanism that provides strategic and operational management of projects in housing and communal services implemented on the basis of the mechanism of the public–private partnership. The relevance of the study of these aspects and the need to develop elements of such a mechanism is determined by both the high social significance and the underfunding of this area of the regional economy during a long time.

Keywords Communal infrastructure systems · Spatial and territorial factors · Regional development · Public utilities

JEL Codes Q48 · Q56 · Q57 · Q53 · O33 · L97

1 Introduction

The housing and utilities sector play an important role in the formation of a comfortable living environment for the population, providing it with basic services that determine the level of quality of life (heat supply, water supply, electricity, gas supply, etc.). Communal infrastructure are also important for the functioning and

S. B. Globa (✉) · D. V. Zyablikov · V. V. Berezovaya
Siberian Federal University, Krasnoyarsk, Russia
e-mail: sgloba@sfu-kras.ru

D. V. Zyablikov
e-mail: DZyablikov@sfu-kras.ru

V. V. Berezovaya
e-mail: VVBerezovaya@sfu-kras.ru

V. N. Ostrovskaya and A. V. Bogoviz (eds.), *Big Data in the GovTech System*, Studies in Big Data 110, https://doi.org/10.1007/978-3-031-04903-3_12

economic activities of industrial, agricultural, commercial enterprises, and other industries [1–3].

Thus, the uninterrupted functioning of housing and communal services enterprises, ensuring the supply of municipal resources of appropriate quality and the safety of communal infrastructure, is one of the most important factors for the sustainable development of territories.

The housing and commercial services sector is the main structural component of the housing maintenance and utilities complex. Comfort and safety of citizens, preservation, proper operation and maintenance of housing stock, uninterrupted operation of enterprises depend on how municipal infrastructure facilities function [4–7].

Municipal infrastructure enterprises provide services and supply resources necessary to meet the vital needs of the population. Depending on the level of need for certain services, we will highlight the following levels:

- *the first level* implies continuous provision of services, utility disruption may lead to emergency situations; it includes the provision of residential, industrial, commercial and public buildings with electricity, heat, water, gas, sanitation; the timing of possible interruptions in the provision of such services is strictly regulated and carefully monitored;
- *the second level* includes services that are necessary for the proper maintenance of buildings, the creation of conditions for living and household activities, but are in demand periodically (for example, cleaning of public areas, elimination of air jams in the heating system, lawn cleaning, removal of snow and ice from roofs, waste removal, etc.);
- *the third level* covers services, the provision of which is carried out on an "as-needed" basis (for example, current and major repairs, replacement of individual elements, replacement of interior lighting lamps, replacement of broken glasses, sprinkling with deicing mixtures, etc.).

The following characteristics are inherent in the housing maintenance and utilities complex:

- territorial isolation, i.e. the connectivity of communal infrastructure facilities with a specific territory and the satisfaction of the needs for public services that are formed on this territory [8–11];
- functional limitation, which is manifested by the presence of a functional limit, determined, on the one hand, by the size and structure of the needs for public services in a certain area in a specific period of time, and, on the other hand, by the capacity and production characteristics of municipal infrastructure facilities [12, 13];
- the dependence of the forms and scale of activity on the spatial and territorial organization of the settlement, zoning of the territory for residential, industrial, infrastructure development [14];
- natural resource dependence of the functioning of municipal infrastructure facilities on the availability and access to water sources, its quality, distance from the energy sources, the need to take into account the terrain, etc.;

- climatic dependence, implying the impact on the functioning of municipal infrastructure facilities of the duration of the period with low temperatures, precipitation patterns, etc.

These characteristics form the territorial features of the functioning of enterprises and affect the organization of their activities.

Krasnoyarsk Krai as a region with a continental climate is characterized by significant average daily temperature differences in winter and summer (in some territories, the greatest difference exceeds 52 °C), as well as day and night, a long period of low temperatures, high snow cover, wind loads on buildings. This largely determines both the lifestyle of the population and the mode of functioning of municipal infrastructure facilities, and also leads to additional costs for the maintenance of infrastructure facilities and maintenance of buildings.

There are six large districts on the territory of the Krasnoyarsk Territory that differ in geographical location and level of socio-economic development: Northern, Angara, Eastern, Western, Central and Southern.

The territories of the Northern microdistrict are located in the harsh arctic climate zone: the average temperature in January is −30°–36 °C, in July + 13 °C, the period with a temperature of more than 10 °C lasts less than 40 days. Polar night is observed in the Taimyr Dolgano-Nenets municipal district, in part of the Evenki municipal district and part of the Turukhansky district. In Norilsk, it lasts 45 days, and you can do without additional lightning for a short time, and in Dudinka—68 days, while 24-h artificial lighting works in settlements. Most of the macro-district is located on permafrost, with buildings built on piles fixed in it. Due to the processes of permafrost melting that have begun due to global warming and improper operation (closing of spaces under houses), the foundations of buildings are destroyed, cracks appear in the load-bearing structures, which even leads to the need to relocate residents. The Northern district is characterized by the fact that almost all types of infrastructure are underdeveloped, starting with transport and energy facilities, necessary to provide it with electric and thermal energy.

The natural and climatic conditions of the Angara district, located in the taiga zone, are also quite harsh and similar to the conditions of the Northern district, the existing transport network and energy infrastructure are insufficiently developed.

The Central, Eastern and Western districts are characterized by more favorable climatic conditions, are the most developed and populated areas of the region, have satisfactory infrastructure provision. The Southern district is distinguished by the longest duration of the frost-free period, the duration of the period with a temperature of more than 10 °C is 110–120 days. The average temperature in January is −18 °C, in July + 20 °C. The average temperature in January is −18 °C, in July + 20 °C.

These characteristics form difficulties in managing and ensuring the functioning of municipal infrastructure facilities, consisting in the need to protect them from low temperatures, soil freezing, as well as ensuring resistance to wind loads, taking into account the height of the snow cover. This requires additional costs and the availability of specialized equipment and reduces the investment attractiveness of municipal infrastructure facilities located in the northern regions. In addition, activities in the territories of small and remote areas are complicated by such additional negative factors as population reduction, shutdown of local production, deteriorated housing stock, which significantly reduces the purchasing power of of the population, reduces the area and the volume of services, increases investment risks. Nevertheless, the municipal infrastructure of these territories is in the most urgent need of investment. So the state's participation in financing such projects should be more significant, taking into account the degree of deterioration of municipal infrastructure facilities and the consumers' ability to pay.

In the larger municipalities and cities, municipal infrastructure facilities most often have a high investment attractiveness. It is determined by a large number of users, higher demand and volume of sales, fast payback, more convenient location, transport accessibility. Therefore, the participation of the state in such projects may increasingly involve the creation of a favorable legal and organizational environment, co-financing part of the costs.

2 Results

In connection with the above, when managing the development of municipal infrastructure, it is necessary to take into account not only industry, but also spatial and territorial features that determine climatic, resource, socio-economic factors that minimize the risks of all subjects of investment activity:

(1) resource efficiency factors:

- functional and moral deterioration of municipal infrastructure facilities;
- the capacity utilization of municipal infrastructure facilities;
- specific consumption of raw materials and other resources required for the proper functioning of of municipal infrastructure facilities;
- shortage of capacity of municipal infrastructure facilities;
- availability of metering devices for communal infrastructure and facilities to which utilities are provided;
- the number of failures and accidents at communal infrastructure facilities;
- the level of losses and unaccounted expenses at municipal infrastructure facilities;
- compliance of the quality of services provided with the established norms and rules;
- quality of public services and utilities;
- the share of utilities replaced annually;
- reliability of the municipal infrastructure facilities.

(2) the level of losses and unaccounted expenses at public infrastructure facilities;

- compliance of the quality of services provided with the established norms and rules;

- quality of public services and utilities;
- the share of utilities replaced annually;
- reliability of the municipal infrastructure facilities.

(3) spatial and territorial factors:

- features of territorial distribution of residents of the territory;
- territorial differences in the level of housing improvement, including due to the different level of their socio-economic situation;
- the volume of the housing stock;
- provision of residents of territories with sources of gas, heat, water supply and sanitation;
- the share of well-maintained housing stock in the territory;
- the share of dilapidated and structurally unsafe housing in the territory;
- duration of the heating season on the territory;
- the level of processing and disposal of municipal solid waste.
- availability of backup sources of water and electricity supply:

(4) climatic factors affecting the quality of utility services:

- the level of minimum, average and maximum annual temperatures;
- wind load level;
- snow cover level;
- amount of precipitation;
- expected duration of the heating period;
- specific natural conditions (permafrost, polar night, etc.).

(5) socio-economic factors (characteristics of utility consumers):

- paying capacity of utility consumers;
- indicators of the standard of living of the population of the territory;
- the share of the population with incomes below the subsistence level;
- service area, the number of utility consumers;
- the level of collection rates among utility consumers.

Based on the assessment of these factors, it is possible to determine the required level of modernization of municipal infrastructure facilities, replacement of worn-out utilities. Moreover, these indicators should be analyzed for each type of utility services and constantly updated.

This will make it possible to position projects for the modernization of municipal infrastructure facilities in order to determine standard groups, management decisions for public authorities on the feasibility of implementing, eliminating or adjusting existing state support measures.

3 Conclusion

The research was carried out within the framework of the research grant of Krasnoyarsk Regional Fund of support scientific and technical activities on the topic "Development of models of financial support for investments in the utilities infrastructure of the region, taking into account the best Russian and world practices and features of the spatial and territorial development of the Krasnoyarsk Territory", No. CF-835, agreement on the procedure for targeted financing No. 226 dated 20.04.2021.

References

1. Infrastructure of Russia: Development index. (2020). Analytical review of InfraOne Research. Moscow, September 2020. https://infraoneresearch.ru/index_id/2020?index2020. Accessed 25 Sept 2021.
2. Katanandov, S. L., & Demin, A. Y. (2021). Problems and prospects of development of the municipal infrastructure management system in the Russian Federation. *Managerial Consulting, 6*, 80–93. https://doi.org/10.22394/1726-1139-2021-6-80-93
3. Oxogoev, A. N. (2013). Communal infrastructure of the region. *Bulletin of ESSUTM, 1*, 122–126.
4. Chanyshev, I. R. (2015). Clustering of the housing and utility complex as one of the directions of its reform. *Financial Analytics: Problems and Solutions, 2*(236), 55–64.
5. Shmakova, M. V. (2013). Factors of spatial development and their consideration in regional strategies. https://elib.bsu.by/bitstream/123456789/88794/1/%D0%A8%D0%BC%D0%B0%D0%BA%D0%BE%D0%B2%D0%B0_%D0%A4%D0%B0%D0%BA%D1%82%D0%BE%D1%80%D1%8B%20%D0%BF%D1%80%D0%BE%D1%81%D1%82%D1%80%D0%B0%D0%BD%D1%81%D1%82%D0%B2%D0%B5%D0%BD%D0%BD%D0%BE%D0%B3%D0%BE%20%D1%80%D0%B0%D0%B7%D0%B2%D0%B8%D1%82%D0%B8%D1%8F.pdf. Accessed 25 Sept 2021.
6. Surnina, N. M., Ilyukhin, A. A., & Ilyukhina, S. V. (2016). Development of social and engineering infrastructure of the region: Essential, institutional, informational aspects. *Proceedings of USUE, 5*(67), 54–65.
7. Sviatokha, N. Y. (2013). Spatial-temporal organization of the region housing sphere: The cluster approach. *Bulletin of Orenburg State University, 8*(157), 140–147.
8. Allin, S., & Walsh, C. (2010). Strategic Spatial Planning in European City-Regions: Parallel Processes or Divergent Trajectories? NIRSA Working Paper Series, 60.
9. Frolova, E. V., Vinichenko, M. V., Kirillov, A. V., Rogach, O. V., & Kabanova, E. E. (2016). Development of social infrastructure in the management practices of local authorities: Trends and factors. *International Journal of Environmental and Science Education, 11*(15), 7421–7430.
10. Yushkova, N. G., Gushchina, E. G., Dontsov, D. G., & Fikhtner, O. A. (2019). Spatial Development Dichotomy: Assessment of The Potential and Implementation of Territorial Systems. Selection and peer-review under responsibility of the Organizing Committee of the conference MTSDT 2019—Modern Tools for Sustainable Development of Territories. Special Topic: Project Management in the Regions of Russia, pp. 792–803.
11. Yushkova, N. G., Gushchina, E. G., Gaponenko, Y. V., Dontsov, D. G., & Gushchin, M. S. (2019). Infrastructural priorities and regularities of spatial development of regional systems. Selection and peer-review under responsibility of the Organizing Committee of the conference CIEDR 2018 The International Scientific and Practical Conference "Contemporary Issues of Economic Development of Russia: Challenges and Opportunities", pp. 474–483.

12. The Decree of the Government of Russian Federation of June 14, 2013 no. 502 “On approval of requirements for programs of complex development of communal infrastructure of settlements and city districts”. https://base.garant.ru/70398922/. Accessed 15 Oct 2021.
13. The Order of the Ministry of Regional Development of the Russian Federation “On the development of programs for the integrated development of communal infrastructure of municipalities” No. 204. of May 06, 2011. https://normativ.kontur.ru/document?moduleId=1&documentId=179076. Accessed 15 Oct 2021.
14. Passport of the Program of integrated development of municipal infrastructure systems of the Krasnoyarsk City district for 2018–2030. http://pravo.admkrsk.ru/Pages/detail.aspx?recordId=32607. Accessed 25 Sept 2021.

P2P Lending as a New Model of Digital Bank

Milyausha K. Khalilova, Vyacheslav A. Davydov, and Shakizada U. Niyazbekova

Abstract To implement the chosen strategy, banks will need their platform, which we will later call a Credit Exchange. At the heart of such a platform is the distributed registry technology and the basic concepts of Basel–2. The description of the tokenization algorithm is given below. An example of the bank's profitability growth as a result of the use of the Credit Exchange is given below. The results of modelling the parameters of the financial market created by the platform are reflected at the end.

Keywords Digital bank · Financial market · Probability · Tokenized

JEL Codes E43 · E59 · G21 · I20

1 Introduction

These factors make it possible to organize interaction in the financial market without intermediaries, which banks traditionally were.

Value

$$\Delta = I\left(1 - \frac{1 + DI\frac{T}{365}}{1 + [D_i]\frac{T}{365}}\right) \quad (1)$$

M. K. Khalilova · S. U. Niyazbekova (✉)
Financial University Under the Government of the Russian Federation, Moscow, Russia
e-mail: shakizada.niyazbekova@gmail.com

V. A. Davydov
National Research University Higher School of Economics, Moscow, Russia

S. U. Niyazbekova
Moscow Witte University, Moscow, Russia

V. N. Ostrovskaya and A. V. Bogoviz (eds.), *Big Data in the GovTech System*, Studies in Big Data 110, https://doi.org/10.1007/978-3-031-04903-3_13

is the Bank's income with a positive sign (or loss with a negative sign) in the procedure of tokenization and placement of packages of credit tokens with number $1 \leq i \leq N$ through the trading platform. Let the credit rate D_i, the validity period of the token T, and also the parameters of the credit PD_i and LGD_i satisfy the inequality

$$M[D_i] = D_i - \left(\frac{365}{T} + D_i\right) PD_i LGD_i > DI \quad (2)$$

Then, when selling each token of a given loan, the bank receives a positive Δ value and can use the funds received to place new loans.

This means that the Bank has underestimated its level of income during the primary placement.

The presence in each package of a large number of tokens of different borrowers allows for a given probability of default PD, a known interest rate D and LGD parameter of each tokenized loan.

2 Literature Review

A basic review of the literature gives an idea of the research topic in the field of the tokenization block of the loan portfolio and the trading platform [4, 5, 12, 14].

3 Methodology

This study was based on the methodology of the basic principles of a systematic approach that creates a rational lending policy and management in the digital economy.

4 Results

Description of the Credit Exchange platform

Tokenization block parameters are set:

I – the amount in rubles for which one token of any of the credits will be realized during the initial placement;

DI – yield in % per annum, which the primary investor will receive in the invested I amount upon initial placement;

The first operation of the algorithm is based on the fact that the received tokens must have the same, predetermined level of the quality parameter. As such a parameter, we will consider the yield that an investor receives acquiring a token. To calculate this parameter, we will assume that the tokenized object has a certain, previously known value C_i after the time T from the moment of tokenization. Let the investor invest in each token the sum of I, and plan to receive the token yield equal to DI after the time T expires [2, 3, 8, 10, 12, 13, 15].

There may be a duplicate.

$T-$ the period in days for which the investor invests in the initial placement;

$n-$ number of tokens in one package.

D_i- the rate of incomes as a percentage of the annual loan i during the period T;

Thus, for loans that are scheduled for payment during the period T, the value D_i will be less than the rate of the loan agreement. On the other hand, if the borrower violated the terms of the contract at the time of tokenization and the Bank owed fines and penalties, the value D_i may be greater than the rate of the loan agreement [4, 5, 7, 11].

The parameter PD_i of the default probability of each token credit with number $1 \leq i \leq N$ during the period T depends on the default probability of the tokenized loan PD_i^* and on what period the bank calculated the parameter PD_i^*.

As a result of the tokenization procedure, tokens are formed that have the same quality characteristics PD_i, LGD_i, D_i we estimate the value of the mathematical expectation of credit yield, taking into account the parameters PD_i, LGD_i, D_i.

Part 1

The n of the highest values $\{z_1^i, z_2^i, \ldots, z_n^i\}$ are selected, from which the z_n^i of token packages are formed. All received blocks are the same and contain one token of each of the n objects with the largest number of tokens. Such a formation is possible, since $z_n^i \leq z_j^i$, $1 \leq j \leq n$. As a result, n of new values are formed: $z_j^{i+1} = z_j^i - z_n^i$, $1 \leq j \leq n$. At least one of the values obtained is zero [5–8].

Thus, the mathematical expectation $M[D_i]$ of the yield in per cent per annum satisfies the following identities:

$$M[D_i] = D_i - \left(\frac{365}{T} + D_i\right) PD_i LGD_i \tag{3}$$

Taking into account the input parameters DI, I, T and the obtained value $M[D_i]$, the tokenization block determines the size of the token p_i for each credit number $1 \leq i \leq N$ by the formula.

$$p_i = I \frac{1 + DI\frac{T}{365}}{1 + [D_i]\frac{T}{365}}, \tag{4}$$

As well as z_i- the number of tokens into which the credit with the number i is divided, calculated by the formula.

$$Z_i = \left\lfloor \frac{E_i}{p_i} \right\rfloor, \tag{5}$$

If the tokenized credit with the number $1 \leq i \leq N$, is not fully repaid during the period T, which will be called the time of the token, then a new procedure of tokenization is carried out for the remaining balance.

The term of placement of the tokenized portfolio on the market. Reduction of terms is achieved by increasing the yield and liquidity of the placed instrument, as well as by confirming the stated level of yield at paying [1, 3, 5, 6, 9, 15].

The number of portfolio sales during the year is set to 1.

Each remaining and re–tokenized loan, the current value of the parameters PD, LGD, EAD, D is again determined, which allows determining the size of the new p token. The received tokens of the new issue are placed again through the trading platform, and the proceeds from the sale of a new token of non–paid loan go to the cumulative credit account and increase the balance of funds r_i. From the definition of R_m and r_i we receive the equality [2, 9, 14]

$$\sum_{i=1}^{N} r_i = \sum_{m=1}^{M} R_m \tag{6}$$

The presence of n independent tokens in each package allows for a given equal probability of default $PD_i = PD$, the same level of default losses $LGD_i = LGD$ and the same interest rate $D_i = D$ of each credit, the token of which is included in the package, to estimate the profitability of the package and the probability of such an event.

An example of the dependence of the rate of return of three packages consisting of 500 tokens, 2,000 tokens and 1,000,000 tokens of profitability of each token $D = 10\%$.

5 Conclusion

Simulation results

Figure 1 shows that as the number of tokens in a package increases, the portfolio yield curve tends to a horizontal straight line passing through a yield point of 8.5%, corresponding to a probability of 0.5 and which is the mathematical expectation of the rate value as a random variable.

For an example in Figure considering $T = 365.0$ $LGD = 45.5\%$, $D = 10\%$, $PD = 3\%$ receive $8.5\% = 10\% - 3\%(1 + 10\%)45.5\%$. We formulate this Statement in general terms.

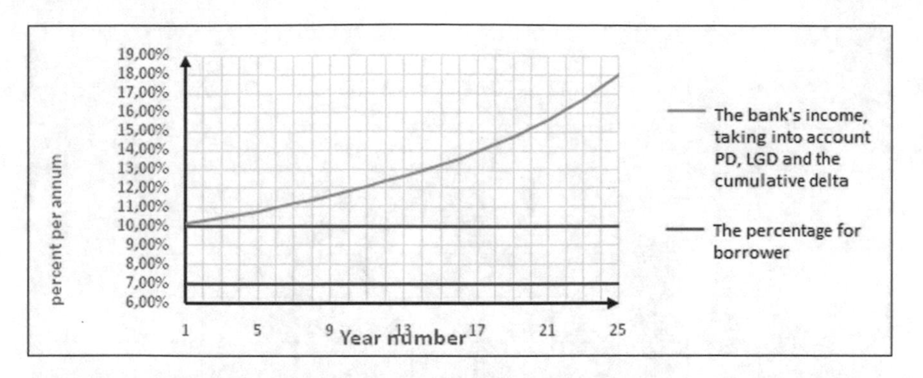

Fig. 1 Comparison of rates of the Bank, the Investor and the Borrower on mortgage loans. *Source* Developed by the authors.

Statement

The distribution of the number of n default tokens of the same size that make up the token portfolio, with the probability of the token default PD, is described by the binomial distribution $B(n, PD)$ with test n parameters and the success probability equal to PD.

The profitability of a portfolio consisting of n tokens, as $n \to \infty$ coincides with the profitability of a portfolio that has a probability of 0.5.

For this distribution the expression $M[B(n, PD)] = nPD$ works.

When the number of default tokens is equal to nPD, these tokens for a period of T will generate an income in the portfolio equal to

$$nPD\left(1 + D\frac{T}{365}\right)(1 - LGD). \tag{7}$$

According to proved statement 1, it is possible for any values of PD, LGD, T and D to find a sufficiently large value of n so that for any arbitrarily small value of δ the condition was met

$$P\left(S \geq D - LGD\left(\frac{365}{T} + D\right)PD\right) \geq 1 - \delta. \tag{8}$$

The resulting inequality allows the tokenization block to ensure, due to the inclusion in the package of a sufficiently large number of tokens n, the required level of reliability for the package buyer and at the same time the guaranteed level of profitability.

The dependence of the income received by market participants on the term for the investment of participants in the asset is shown in Fig. 2. The dependence of the income on the asset holding period is well approximated by a logarithmic function. The graph shows the parameters of such a function, as well as the resulting quality of the approximation.

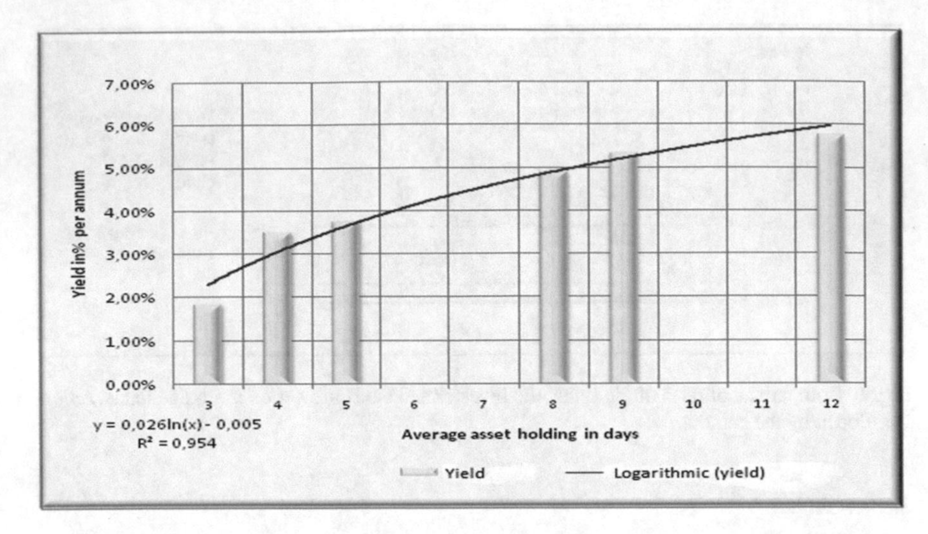

Fig. 2 Income on investment of Participants depending on the period. *Source* Developed by the authors

It should be noted that the proposed system allows the participants to receive a positive income (2% per annum) at the minimum asset holding periods (3 days).

Trading platform block

We will call such a registry, the registry of the trading platform.

During the primary placement, the *M* of formed packages of tokens arrives at the entrance of the trading platform block. By the yield, they have the same characteristics for the primary placement.

References

1. Baidalinova, A.S., Baigireyeva, Z., Myrkanova, A. (2021). Household Food Security in Kazakhstan. In: Popkova, E.G., Sergi, B.S. (Eds). "Smart Technologies" for Society, State and Economy. ISC 2020. Lecture Notes in Networks and Systems, 155. Cham: Springer. https://doi.org/10.1007/978-3-030-59126-7_13.
2. Baranov, D.N. (2018). The essence and content of the category "Digital Economy". Bulletin of the Witte Moscow University. Series 1: Economics and Management, 2(25), 15–23. https://doi.org/10.21777/2587-554X-2018-2-15-23.
3. Brodunov, A.N., Ushakov, V.Ya. (2015). Justification of financial decisions in conditions of uncertainty. Bulletin of the Witte Moscow University. Series 1: Economics and Management, 1(12), 30–36.
4. Davydov, V., & Khalilova, M. (2018). The business model of creating a digital platform for the tokenization of assets on financial. *IOP Conference Ser. Mater. Sci. Eng, 497*, 012069.
5. Gavrilova, E. N. (2018). Credit history bureau: formation, evaluation of efficiency and ways of improvement. *Bulletin of the Witte Moscow University. Series 1: Economics and Management, 4*(27), 34–42. https://doi.org/10.21777/2587-554X-2018-4-34-42.

6. Gavrilova, E. N. (2019). Investment banking as a direction of banking activity: the essence, features and problems of development. *Bulletin of the Witte Moscow University. Series 1: Economics and Management, 4*(31), 81–86. https://doi.org/10.21777/2587-554X-2019-4-81-86.
7. Grekov, I. E., & Blokhina, T.K. (2016). The influence of macroeconomic factors on the dynamics of stock exchange in the Republic of Kazakhstan. *The Economy of the Region, 12*(4), 1263–1273. https://doi.org/10.17059/2016-4-26
8. Ivanova, O. S., Suleimenova, B., Yerzhanova, S. K., & Berstembayeva, R. K. (2021). Oil and gas investment opportunities for companies in modern conditions. In E. G. Popkova, V. N. Ostrovskaya, A. V. Bogoviz (Eds.), *Socio-economic Systems: Paradigms for the Future.* Studies in Systems, Decision and Control, (Vol. 314). Cham: Springer. https://doi.org/10.1007/978-3-030-56433-9_70.
9. Kunanbaeva, K., Madiiarova, D., & Sagyndykova, R. (2020). Development of integration processes in the mining and metals sector in Russia and Kazakhstan under business transformation. In *E3S Web of Conferences*, (Vol. 210). https://doi.org/10.1051/e3sconf/202021013032. ISSN: 25550403.
10. Kurmankulova, R. Z., Anzorova, S. P., Goigova, M. G., & Yessymkhanova, Z. K. (2021) Digital transformation of government procurement on the level of state governance. In: E. G. Popkova, V. N. Ostrovskaya, A. V. Bogoviz (Eds.), *Socio–economic systems: Paradigms for the future.* Studies in Systems, Decision and Control, (Vol. 314). Cham: Springer. https://doi.org/10.1007/978-3-030-56433-9_69.
11. Maisigova, L. A., Isayeva, B. K., & Dzholdosheva, T. Y. (2021). Features of relations between government authorities, business, and civil society in the digital economy. In E. G. Popkova, V. N. Ostrovskaya, A. V. Bogoviz (Eds.), *Socio–economic systems: Paradigms for the future.* Studies in Systems, Decision and Control, (Vol. 314). Cham: Springer. https://doi.org/10.1007/978-3-030-56433-9_144.
12. Moldashbayeva, L. P., Zhumatayeva, B. A., Mezentseva, T. M., & Shirshova, L. V. (2021). Digital economy development as an important factor for the country's economic growth. In E. G. Popkova, V. N. Ostrovskaya, A. V. Bogoviz (Eds.), *Socio–economic systems: Paradigms for the future.* Studies in Systems, Decision and Control, (Vol. 314). Cham: Springer. https://doi.org/10.1007/978-3-030-56433-9_38.
13. Nurpeisova, A. A., Smailova, L. K., Akimova, B. Z., & Borisova, E. V. (2021). Condition and prospects of innovative development of the economy in Kazakhstan. In E. G. Popkova, V. N. Ostrovskaya, A. V. Bogoviz (Eds.), *Socio–economic systems: Paradigms for the future.* Studies in Systems, Decision and Control, (Vol. 314). Cham: Springer. https://doi.org/10.1007/978-3-030-56433-9_184.
14. Patashkova, Y., Kerimkhulle, S., Serikova, M., & Troyanskaya, M. (2021). Dynamics of Bitcoin trading on the Binance cryptocurrency exchange. *Economic Annals-XXI, 187*(1–2), 177–188.
15. Rakhimova, S., Kunanbayeva, K., Goncharenko, L., & Pigurin, A. (2019). A balanced system of indicators for the assessment of innovative construction projects efficiency. In *E3S Web of Conferences*, (Vol. 110, p. 02154). https://doi.org/10.1051/e3sconf/201911002154. ISSN: 25550403.

Innovations in Accounting and Analytical Support in the Construction of Automated Integrated Systems

Svetlana A. Chernyavskaya, Nina V. Khodarinova, Tatyana E. Glushchenko, Elena B. Sheludko, and Anna P. Anufrieva

Abstract Accounting of current assets must be aimed at improving the efficiency of its utilization, the economy of consumption, and improved planning of current assets. Thus, efficient management of current assets is eventually responsible for the profitability of a business entity, which determines the timeliness of the research topic. It is against this background that we need to develop methods for improving the efficiency of utilization of current assets of business entities. The subject of research in this paper lies in the issues of identifying the major lines of development of accounting and the analysis of the efficiency of utilization of current assets. The usefulness of the paper lies in the possibility to use research results to improve the accounting methodology and analyse current assets in agricultural organizations. The scientific novelty of research results consists in the fact that unique proposals have been developed that are aimed at improving accounting and the efficiency of current capital utilization in JSC Rossiya of the Kanevskoy District of the Krasnodar Territory.

Keywords Accounting · Automation · Control · Analysis · Current assets · Agriculture

JEL Code M410

S. A. Chernyavskaya (✉)
Kuban State Agrarian University Named After I.T. Trubilin, Krasnodar, Russia
e-mail: docsveta17@gmail.com

N. V. Khodarinova · T. E. Glushchenko
Russian University of Cooperation, Mytishchi, Russia

E. B. Sheludko · A. P. Anufrieva
Kuban State Technological University, Krasnodar, Russia

V. N. Ostrovskaya and A. V. Bogoviz (eds.), *Big Data in the GovTech System*,
Studies in Big Data 110, https://doi.org/10.1007/978-3-031-04903-3_14

1 Introduction

Own funds form the core of activity and are formed from assets of the organization as such, belonging to it on the right of ownership. Assets from this source constitute the financial stability of the economic entity. However, to arrange for the efficient management of current assets to the greatest possible extent, own funds alone turn out to be insufficient; as a result, the business entity further forms its current capital at the expense of borrowed and employed assets. The source of borrowed current assets includes short-term credits and loans, at the expense of assets of which the business entity satisfies its needs for current capital. Borrowed assets are those current assets that are not owned by an enterprise but are included in its turnover. These may include credit debts, advanced payments by buyers for products, amounts paid by buyers as a pledge for reusable containers, etc.

2 Materials and Methods

A reasonable correlation between own current assets and borrowed current assets and employed current assets is important as well. This correlation determines the level of financial stability of a business entity. If the total amount of borrowed and employed current assets is much higher than the total amount of own current assets, this brings grave risks to the business entity.

Before the study of the financial and operating activities of leading business entities, it is expedient to take a look at the current situation in the agricultural sector of the Krasnodar Territory [1].

The number of business entities in the sector and its dynamics constitutes an important index of the sector's development. The ratio of unprofitable business entities is of particular importance as well. No significant changes in the number of agricultural organizations for the period under study have been observed (deviation from mean value is lower than 3.5%). That said, the number of unprofitable business entities has an upward trend: in 2019, their number increased by 50% compared to 2015. Even though the number of unprofitable business entities has decreased by 22.08% in 2019 compared to 2018, their number and proportion in the total number of agricultural organizations are still high.

To establish the reasons for these changes, the analysis of the debt load of agricultural organizations was performed (Fig. 1).

It can be seen that the volumes of collected revenue and liabilities for the entire research period are roughly equal. This suggests that the level of debt load of agricultural producers, fairly high as it is, is not decreasing [2]. This situation results in the shortage of own current assets in economic entities of the sector. To determine the influence of the price factor on the increase in volumes of agricultural products, the analysis of changes in average prices for agricultural products was performed.

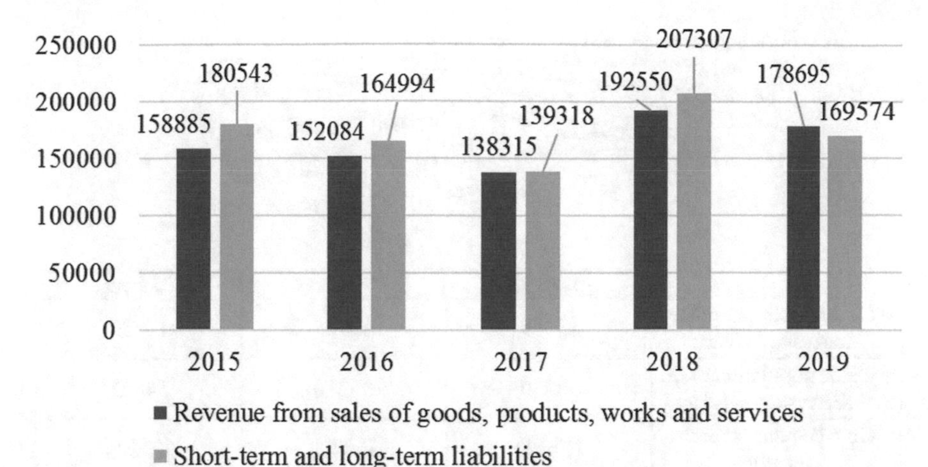

Fig. 1 Correlation between revenue and liabilities in agricultural organizations of the Krasnodar Territory, million roubles. *Source* Compiled by the author

We may conclude that prices for certain types of crop products are increasing at a fast pace. For example, the prices for 1 ton of wheat has increased by 27.89% or by 2,433 roubles for the recent 3 years. That said, average prices for livestock products are slightly increasing or even decreasing at all. Thus, the average selling price for unpasteurized milk of cattle has decreased by 0.2% or by 54 roubles per ton in 2019 compared to 2017.

Changes in prices for certain types of agricultural products should be compared with changes in prices for goods and services, purchased by agricultural organizations [3]. The increase in prices for certain goods that are purchased by agricultural organizations exceeds the rate of growth for certain types of products. This, in turn, is manifested in the significant rise in prices for energy sources. As a result, a disparity of prices occurs, which leads to the "shrinkage" of the agricultural sector. With such a significant rise in prices for bought-in goods, production of certain types of agricultural products (especially dairy breeding) becomes less profitable, and even unprofitable for some agricultural producers [4].

As a next step, consideration has been given to the structure and dynamics of crop areas of certain crop plants. The crop areas of crop plants both in the Kanevskoy District and in the entire Krasnodar Territory are still yet decreasing. Thus, in 2019, the total amount of crop areas in the Krasnodar Territory decreased by 2.09% or by 52 thousand hectares compared to 2015; further, it has decreased by 0.13% or by 3.1 thousand hectares compared to 2018. In addition to the above, against the background of reduced crop areas of certain crop plants in the entire Krasnodar Territory, they have an upward trend in the Kanevskoy District.

As a next step, we performed the analysis of the availability and dynamics of the livestock population in the entire Krasnodar Territory and the Kanevskoy District. This information is presented in graphic form in Fig. 2.

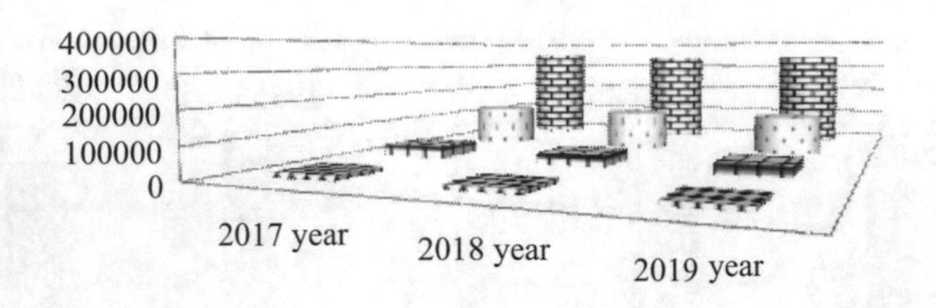

	2017 year	2018 year	2019 year
Cow population in the Kanevskoy District	15509	13705	13659
Cattle population in the Kanevskoy District	38270	35310	34802
Cow population in the territory	131300	128400	127100
Cattle population in the territory	350800	336700	338100

Fig. 2 Availability and dynamics of cattle population and population of cows for the Krasnodar Territory and the Kanevskoy District, animal units. *Source* Compiled by the author

Figure 2 suggests that cattle population and cow population has been decreasing both in the Kanevskoy District and in the entire Krasnodar Territory. Thus, in 2019, the cow population throughout the territory has decreased by 3.20% or by 4200 animal units compared to 2017; compared to 2018, it has decreased by 1.01% or by 1300 animal units. This being said, a decrease in the population can be observed in the Kanevskoy District as well: in 2019, their population has decreased by 11.93% or by 1850 animal units compared to 2017; compared to 2018, it has decreased by 0.34% or by 46 animal units. Such changes are due to the influence of several external negative factors: increase in prices for purchased goods and services; low government subsidies for milk production; realization of the bulk of profit by sales organizations rather than by immediate producers, etc.

As can be seen from the above, the upward trend in the ratio of unprofitable agricultural organizations, high debt load, disparity of prices for goods purchased and products sold, a decrease in crop areas and cattle population prevent from making a positive evaluation in the agricultural sector of the Krasnodar Territory [5].

Table 1 presents the profitability indices in leading business entities.

Return on assets in JSC Rossiya in 2019 was 8.49%, which is lower than the index for 2015 by 3.42%, but higher than the index for 2018 by 2.27%. In JSC Urozhai, this index was 10.32%, having decreased by 11.12% compared to 2015, and by 5.31% compared to 2018. In LLC Kuban, it was 10.03% in 2019, having decreased by 12.05% compared to 2015, and having increased by 1.35% compared to 2018. The decrease in this index for the long period suggests the decrease in the volume of profit earned before tax against the background of a decrease in the cost of aggregate capital, which describes the performance results of these economic entities

Table 1 Profitability indices in agricultural enterprises, %

Index	2015	2016	2017	2018	2019	Absolute deviation (+, −), 2019, as compared to	
						2015	2018
Return on assets							
JSC Rossiya	11.91	11.63	10.49	6.22	8.49	−3.42	2.27
JSC Urozhai	21.44	5.44	30.46	15.63	10.32	−11.12	−5.31
LLC Kuban	22.08	23.92	19.06	8.67	10.03	−12.05	1.35
Return on equity							
JSC Rossiya	23.08	19.56	11.23	7.85	8.79	−14.28	0.94
JSC Urozhai	39.58	36.82	17.84	13.47	15.79	−23.79	2.32
LLC Kuban	24.80	19.67	9.29	12.58	24.28	−0.51	11.70
Profitability of products sold							
JSC Rossiya	35.82	30.79	30.01	29.97	30.27	−5.55	0.30
JSC Urozhai	36.69	52.00	44.80	41.03	35.77	−0.92	−5.26
LLC Kuban	38.01	30.83	19.07	15.56	61.35	23.34	45.78
Margin on sales							
JSC Rossiya	29.23	39.27	31.23	25.71	27.67	1.56	1.96
JSC Urozhai	40.65	11.21	84.71	37.76	18.27	−22.38	−19.49
LLC Kuban	36.45	33.36	32.25	19.94	10.33	−26.13	−9.62
Net margin of sales							
JSC Rossiya	21.59	23.62	16.19	10.52	11.50	−10.09	0.97
JSC Urozhai	19.99	48.42	49.30	41.46	31.33	11.34	−10.13
LLC Kuban	32.77	25.07	13.52	25.76	50.71	17.93	24.95
Net margin of production							
JSC Rossiya	33.64	34.13	23.13	15.03	16.49	−17.15	1.46
JSC Urozhai	31.58	100.88	89.32	70.32	48.78	17.20	−21.54
LLC Kuban	52.87	36.24	16.70	30.51	131.19	78.32	100.69

Source Compiled by the author

negatively. That said, the net margin of sales in JSC Rossiya in 2019 was 11.50%, which is less than the index for 2015 by 10.09% and is higher than the index for 2018 by 0.97%. In JSC Urozhai, this index was 31.33%, which is higher than the index for 2015 by 11.34% but is lower than the index for 2018 by 10.13%. In LLC Kuban, this index in 2019 was 50.71, having significantly increased by 17.93% compared to 2015; having increased by 24.95% compared to 2018.

This methodology is based on the fact that financial stability is determined through the construction of a three-component model with the use of indices that were calculated above [6]. If the lack of payment is found, a value of “1” is assigned, and if the lack of payment is found, a value of “0” is assigned (Table 2).

Table 2 Results of the analysis of the type of financial situation in business entities under study

Conventional notation	2015		2016		2017		2018		2019	
	Amount, thousand roubles	Assigned value	Amount, thousand roubles	Assigned value	Amount, thousand roubles	Assigned value	Amount, thousand roubles	Assigned value	Amount, thousand roubles	Assigned value
Surplus (+) *or lack* (−) *of own current assets* (*ΔCOC*)										
JSC Rossiya	−255,118	0	−253,442	0	−27,9013	0	−318,600	0	−448,055	0
JSC Urozhai	−357,830	0	302,938	1	480,535	1	404,850	1	542,883	1
LLC Kuban	−4,287	0	−12,572	0	−19,753	0	50,516	1	32,567	1
Surplus (+) *or lack* (−) *of own sources of funds and long-term sources of borrowed funds for inventory carrying and costing* (*ΔСД*)										
JSC Rossiya	−188,678	0	−195,773	0	−203,203	0	−241,490	0	−272,532	0
JSC Urozhai	−357,830	0	302,938	1	480,535	1	404,850	1	542,883	1
LLC Kuban	15,713	1	7,428	1	247	1	50,516	1	32,567	1
Surplus (+) *or lack* (−) *of the total amount of major sources of funds for inventory carrying* (*ΔОИ*)										
JSC Rossiya	−127,398	0	−79,773	0	−83,229	0	−108,241	0	−107,506	0
JSC Urozhai	284,128	1	347,753	1	480,535	1	404,850	1	882,883	1

(continued)

Table 2 (continued)

Conventional notation	2015		2016		2017		2018		2019	
	Amount, thousand roubles	Assigned value	Amount, thousand roubles	Assigned value	Amount, thousand roubles	Assigned value	Amount, thousand roubles	Assigned value	Amount, thousand roubles	Assigned value
LLC Kuban	15,713	1	15,986	1	33,131	1	56,301	1	32,567	1
Three-component index of the type of financial situation										
JSC Rossiya	{0; 0; 0}		{0; 0; 0}		{0; 0; 0}		{0; 0; 0}		{0; 0; 0}	
JSC Urozhai	{0; 0; 1}		{1; 1; 1}		{1; 1; 1}		{1; 1; 1}		{1; 1; 1}	
LLC Kuban	{0; 1; 1}		{0; 1; 1}		{0; 1; 1}		{1; 1; 1}		{1; 1; 1}	
Type of financial situation										
JSC Rossiya	Crisis		Crisis		Crisis		Crisis		Crisis	
JSC Urozhai	Unstable		Excellent		Excellent		Excellent		Excellent	
LLC Kuban	Stable		Stable		Stable		Excellent		Excellent	

Source Compiled by the author

As can be seen from the above, the crisis financial situation was found in JSC Rossiya throughout the period of research. The top management of this economic entity should promptly take measures to improve financial stability. In JSC Urozhai, an excellent financial situation has been found throughout the research period (except the index for 2015). The actions of managers of this business entity must be aimed at preserving financial stability (for example, performing diversification of production, making profitable financial investments). In LLC Kuban, financial standing has greatly improved for the period under study.

3 Results

To improve accounting and the efficiency of current capital utilization in JSC Rossiya, we have recommended the following activities:

(a) using advanced features of 1S 8.3 "Integrated automation" software, arrange for the management of the flow of materials and finished products in the storages "Storage of a threshing farm" and "Feed storage of farm 2", "Feed storage of farm 3" with an itemization of storage locations. For example, for the storage: "Storage of a threshing farm": "Barn 1"; "Barn 2"; "Outdoor area"; "Feed storage of farm 2": "Bunker silo 1"; "Loft barn"; "Feed processing building". To achieve this goal, the "Storage location" is indicated in all documents on the flow of materials and finished products, and accounting records are formed with an additional itemization of storage locations.

In addition to the above, the features of 1S "Integrated automation" 8.3 software make it possible to generate model reports with necessary parameters at the level of itemization; for example, for account 10 "Materials" or account 43 "Finished products": storage; stock items; storage locations. In addition, it is possible to use additional itemization for the "Units" directory, where necessary. Model reports: "Account analysis", "Turnover balance sheet for the account" [7].

In the management of real-time accounting of the flow of feeds with the use of cattle-feeders with electric balances and the ProFeed software, which enables the transfer of information on loaded feeds depending on their types based on the types of stock items. For example:

- food type: "Hay";
- stock item type: "Alfalfa of prior years—hay"

We have recommended additionally indicating the "Storage location" while loading information from the ProFeed software to the "Agriculture" block of the "Feeds writing-off" document, and the following accounting records will be formed then:

Debit side 20 "Main production unit" subsidiary account 2 "Livestock products":

- "Unit" directory: "Farm 2";

- "Stock item group" directory: "Forage-fed cows";
- "Account cost" directory: "Own feeds"

Credit side 10 "Materials", subsidiary account 14 "Feeds":

- "Stock items" directory: "Alfalfa of prior years—hay";
- "Storage" directory: "Feed storage of farm 2";
- "Storage location: Loft barn" directory.

This itemization of storage locations not only in quantitative but also in value terms allows increasing the control over preservation of stocks and finished products, taking into account both carry-over feedstocks of prior years (monitoring their availability and feeding order according to shelf life and nutritional value of feeds); in case of long-term storage of market products (grains, sunflower seeds) in sealed warehouses until they are sold, improving the speed and reliability of the inventory check, taking into account the complexity and high cost of grain reweighting.

(b) further, we have recommended that the audit committee shall conduct a monthly analysis of humidity indices for the remainder of grain and sunflower seeds in storage to justify, in case of deficiencies, write-offs within the limits of natural losses and reduced grain humidity since harvesting (July) to disposal (May of the following year).

(c) to reduce the cost of rough and succulent feeds under the conditions of insufficient rainfall in the Northern zone of the Krasnodar Territory (location of JSC Rossiya), we have recommended using the drip subsoil irrigation system. This will greatly increase the yield of forage crops, thereby reducing the ratio in the structure of forage crop rotation, freeing up areas for commercial crops, which will contribute to an increase in revenue, improving the efficiency, solvency and financial stability of a particular agricultural producer.

(d) given the existing technique of storing rough and succulent feeds (silage, haylage) we have pointed out certain disadvantages such as ingress of moisture, deterioration in qualitative characteristics and nutritional value of feeds, spoilage). Apart from building new expensive permanent bunker silos, we have recommended creating a "green forage chain" during the grazing period from May 15 till October 15. The use of uniform cow feeding has several disadvantages: ration cost increase due to the need for the use of feed additives to achieve its balance; ration cost increase at the expense of losses during storage and reduced nutritional value of feeds; unquestionable benefits of the nutrient availability of vitamins in fresh herbage compared to hay, silage and haylage; lower quality of store-bought feed additives (in the absence of own laboratories) compared to foreign countries (Israel, United States, etc.) which recommend a uniform feeding ration. Thus, under the current conditions of operation of dairy breeding in the Northern zone of the Krasnodar Territory, the use of "green forage chain" while feeding dairy cows will contribute to improving the productiveness of farm livestock; increased breeding indices (calf crop per 100 cows); increased life expectancy of cows; milk and breed cost reduction.

4 Conclusion

Judging from the research performed, it is obvious that accounting and the analysis of the efficiency of current capital utilization are highly important in financial and operating activities in agricultural organizations, and the issue of their improvement should be among the priority issues. The following practical proposals which contribute to the development of accounting and the analysis of the efficiency of current capital utilization have been made following the research results: it is necessary to reequip the machine and tractor fleet to optimize material costs and improve the efficiency of utilization of current assets (spare parts) in JSC Rossiya; it has been determined that to improve the efficiency of current capital utilization, it is necessary to do the following: apply resource-saving technologies, increase control over the consumption of materials, aim for the cost reduction of products (judgmental factor) and, accordingly, increase profit on sales; then this will greatly change the efficiency of current capital utilization; aim at increasing the product yield at the expense of intensive factor (increase in productivity of basic crops and productiveness of farm livestock (milk yield), etc.).

References

1. Bzhasso, A. A. (2014). Key objectives of anti-crisis regional economies management. *Bulletin of the Adyghe State University. Series 5: Economy*, *2*(141), 30–35.
2. Kartashov, K. A., Isachkova, L. N., Sotskaya, T. V., & Kunakovskaya, I. A. (2018). Digital economy as a basis of the modern world or new problems for the Russian society. *Bulletin of the Adyghe State University. Series 5: Economy*, *4*(230), 167–173.
3. Hodarinova, N. V. (2017). Organization and evaluation of the effectiveness of internal and external control of modern agricultural organizations. *Journal of Economy and Entrepreneurship, 4–1*(81), 536–547.
4. Cozzani, V., & Zanelli, S. (2007). An approach to the assessment of domino accidents hazard in quantitative area risk analysis. *Journal of Hazardous Materials, 130*, 24–50.
5. Stein, J., Usher, S., LaGatutta, D., & Youngen, J. (2001). A comparables approach to measuring cashflow-at-risk for non-financial firms. *Journal of Applied Corporate Finance, 13*(4), 100–109.
6. Frantsisko, O. Y., Ternavshchenko, K. O., Molchan A. S., Ostaev, G. Y., Ovcharenko, N. A., & Balashova, I. V. (2020). Formation of an integrated system for monitoring the food security of the region. *Amazonia Investiga [Amazonia Bulletin]*, *9*(25), 59–70.
7. Chernyavskaya, S. A., Berkaeva, A. K., lyanova, S. A., Kashukoev, M. V., & Misakov, V. S. (2020). Regional and sectoral system for integrated assessment and green supply chain management of natural resources. *International Journal of Supply Chain Management*, *9*(2), 714–718.

Digitalization of the Strategic Management Systems of the Oil and Gas Industry Enterprises

Abror S. Kucharov, Azizjon B. Bobojonov, Elvira A. Kamalova, Dinora N. K. Ishmanova, and Bakhtiyor J. Ishmukhamedov

Abstract This article examines the theoretical foundations of infrastructures, digitalization of the strategic management system, the economic essence, theoretically analyzes the current state, the need and importance of infrastructures at oil and gas enterprises. Also, the institutional features and principles of infrastructure management of oil and gas enterprises, the features of the application of the best practices of foreign countries in improving the state management of the oil and gas system have been investigated.

Keywords Digital economy · Digitalization · Digital transformation · Strategic management systems · Oil and gas industry · National economy · Foreign economic activity

JEL Code L95

1 Introduction

In conditions of strong competition, it is advisable to develop ways to implement a strategy for achieving a target competitive position when substantiating the main directions of a strategy for the development of oil and gas enterprises. Improving the mechanisms of strategic management of innovation processes in monopolistic sectors and further improving the competitive environment, digitalizing the strategic management system, increasing the efficiency of management of oil and gas enterprises, reducing state participation in the economy is one of the priorities areas.

"Infrastructure" from the Latin word (infrastructure) means "outside the structure". From an economic point of view, the following explanation is more suitable for the essence of infrastructure: "it is a set of specific labour processes in the creation

A. S. Kucharov (✉) · A. B. Bobojonov · E. A. Kamalova · D. N. K. Ishmanova · B. J. Ishmukhamedov
Tashkent State University of Economics, Tashkent, Uzbekistan
e-mail: abrork1967@gmail.com

V. N. Ostrovskaya and A. V. Bogoviz (eds.), *Big Data in the GovTech System*, Studies in Big Data 110, https://doi.org/10.1007/978-3-031-04903-3_15

of goods and services that ensure the exchange of activities in the process of human life and social production." Taking into account the above, we have expressed the economic classification of the term infrastructure as follows: "Infra" -under, lower, "structure" -system, structure. The basis, the foundation for the development of sectors of the economy is a complex of structures [5].

In subsequent years, the infrastructure developed at a rapid pace. This can be attributed to several factors. In particular, the rate of growth of production outstrips the development of infrastructure, which also has an impact on the development of the economy. There is widespread research on the impact of infrastructure on the economic growth and development of countries. However, most researchers get around the problem of identifying a clear understanding of the object under study and its criteria. Canning and Pedroni, analyzing the impact of infrastructure on economic growth based on an econometric model, found that investment in infrastructure is to a certain extent profitable [4].

In [3] opinion there was a significant effect from investments in infrastructure. It means a 1% increase in infrastructure investment results in increased production by 0.39% in the private sector of the economy.

2 Methodology

An important condition for the dynamic development of the Republic of Uzbekistan is the accelerated introduction of modern innovative technologies in the oil and gas sector of the economy [6]. The Strategy of Actions for the Development of the Republic of Uzbekistan for 2017–2021 is of great importance in the development strategy and effective management of joint-stock companies. In the action strategy, as priority areas for the development and liberalization of the economy, "the introduction of modern standards, principles and methods of corporate governance, the digitalization of the strategic management system, the role of shareholders in the strategic management of enterprises in the development of joint-stock companies and the organization of effective management" is of great importance [2].

The stages of this reform included partial or full privatization on behalf of labour communities, banks, investment companies and foreign investors based on investment commitments through the sale of blocks of shares. 609 state assets were privatized with a total value of 174.8 billion soums and 1.1 million dollars, as well as with an investment obligation in the amount of 580.8 billion soums and 30.2 million dollars, including 275 objects, sold at zero cost on a competitive basis. More than 13 thousand new jobs have been created at these facilities.

In 2019, the gas transportation system of the Republic of Uzbekistan consisted of more than 13.6 thousand kilometres of main gas pipelines. The supply of natural gas to consumers of the Republic of Uzbekistan is carried out by the Underground Gas Storage Administration in the East, North and South directions and main gas pipelines. Uzbekneftegaz JSC enterprises, today there are 25 gas pumping stations, 252 gas pumping units, 393 gas distribution stations, gas distribution points—101,317 units, main gas pipelines—13,250.2 km, high-pressure gas pipelines—13,736 units, medium-pressure gas pipelines—31,651 km, low-pressure gas pipelines—85,203 km, available free gas fields—210 units, oil fields—125 units.

According to the authors, reliable and uninterrupted operation of main gas infrastructure facilities is ensured by the followings:

- continuous monitoring of the state of the branch part of gas pipelines, diagnostics of the technical condition of pipelines, cleaning of the inner part of gas pipelines and repairs, maintenance of gas pipelines based on the results of flaw detection of the inner part of the pipe;
- diagnostics, technical repair and maintenance of gas filling stations, communications and technological equipment of internal premises;
- modern reconstruction and modernization of gas transmission system facilities, digitalization of the strategic management system;
- continuous training and advanced training of technical specialists.

From the Table 1, one can draw such a clear conclusion from the indicators that the use of natural gas in energy devices (power plants and greenhouses) that produce electricity and heat energy in Germany has the least negative impact on the environment. After it is burned, the solid parts of sulfur oxide (liquid and solid, which are released during combustion) are almost not released (Table 1).

Lukoil is one of the world's largest vertically integrated companies engaged in the extraction and processing of oil and gas, the production of petroleum products and petrochemical products. The company occupies a leading position in the Russian and world markets in its main areas of activity.

The company's goal is to channel the energy of natural resources for human prosperity, promote long-term economic growth, social stability in the regions of its operation, create opportunities for prosperity and development, maintain a favourable environment and rational use of natural resources. After gaining independence, Uzbekistan entered the twenty-first century as the top ten countries among more than 90 natural gas producing countries in the world.

Table 1 Air pollution of a farm from the combustion of various fuels in Germany

Source of pollution	Contaminants							
	carbon		Carbon monoxide		Nitric oxide		Sulfur oxide	
	oil, coal	gas	oil, coal	gas	oil, coal	gas	oil, coal	gas
Transport	87.6	0	87.2	0	50.0	0	3.0	0
Power plant	1.2	0	0.4	0	30.0	5	63	0
Communal services	9.2	0	21.0	0	4.0	1	14	0
Industry	2	0	0.4	0	6.0	4	20	0
Total	100	0	100	0	90	10	100	0

Source Authors' based on [7]

Among the 20 countries that produce the largest amount of natural gas in the world, in terms of annual natural gas production (in billion cubic meters), Uzbekistan ranks 12th with a total value of 65 billion cubic meters and its share in the world market is 1.8%.

Following the Action Strategy for the five priority areas of development of the Republic of Uzbekistan, some changes were also made to the oil and gas industry. In particular, the Decree of the President of the Republic of Uzbekistan "On measures to improve the oil and gas sector management system" dated June 30, 2017, improved the activities of the Joint Stock Company "Uzbekneftegaz" [1]. In 2019, the plan was implemented for the production of oil and gas condensate, the production of liquefied natural gas. However, many problems have accumulated in the industry over the years. Since the reserves are not increasing, the forecast for the production of natural gas and the production of petroleum products in 2018 has not been fulfilled sufficiently. In terms of drilling volumes in the course of geological exploration, the forecast was fulfilled by 77%. In addition, the forecast for production drilling was fulfilled by 49% and the forecast for the number of completed wells was fulfilled by 53%. Capacities for natural gas production, modernization and repair of distribution networks are also insufficient. As a result, 6% of this natural resource, starting from gas production, is lost before reaching the consumer.

In the practice of the developed countries of the world, a lot of work is being done to study the theoretical and methodological aspects of the activities of corporations, the choice of strategies for their management, a comprehensive study of the problems of managing joint-stock companies based on public–private partnerships. To introduce and develop an effective management system for the oil and gas industry, increase its competitiveness and investment attractiveness, as well as the Action Strategy for five priority areas of development of the Republic of Uzbekistan for 2017–2021 following the tasks set out in the concept of administrative reforms in the Republic of Uzbekistan:

The priority areas for further development of the fuel and energy sector of the Republic of Uzbekistan are followings:

- maintaining a unified energy policy of the country aimed at ensuring fuel and energy security, at meeting the constantly growing demand of the economy and the population for energy resources;
- state regulation in the oil and gas sector, a clear limitation of the functions of economic activity, improvement of the regulatory and institutional framework for social and public–private partnerships, development of clear market structures for the implementation of tariff policy and, on this basis, promotion of the principles of a healthy competitive environment;
- creating conditions for actively attracting investments in the construction of infrastructure facilities, as well as the modernization of industrial enterprises, technical and technological re-equipment, primarily foreign direct investment;
- widespread use of modern means of automation of technological processes, systems for recording production volumes, supply and consumption of energy resources at enterprises of the energy industry;

- introduction of modern methods and target indicators (quality management, indicative planning) aimed at coordinating the management system of sectoral enterprises, their structures and divisions, achieving accurate results of the organization of work, as well as the introduction of a unified legal and technical regulation in the field of the fuel and energy industry in the republic;

The fact that the state commission for tendering in the sale of state property has been granted the right to amend the conditions and methods of selling state-owned shares packages of business entities' stocks to foreign investors indicates that the country is using foreign experience.

Discussed a draft resolution of the President of the Republic of Uzbekistan "On additional measures to reduce the state share in the economy from the number of enterprises fully or partially held in state blocks of shares (stakes) based on introducing the principles of the strategy of ownership, management and reform of enterprises with state participation of the Republic of Uzbekistan in 2020–2025. In turn, this testifies to the correspondence of the scientific theory proposed by us to this draft resolution (Table 2).

Ten state-owned enterprises with state participation, in which state stock share packages in authorized capital are sold in full to the private sector on the territory of our country while maintaining the management of oil and gas enterprises of strategic importance, which will lead to the economic recovery of enterprises with financial instability in the industry.

According to the analysis in the draft Resolution of the President of the Republic of Uzbekistan "On additional measures to reduce the share of the state in the economy" in the list of enterprises with the participation of the state, where the state block of shares (share) is fully or partially retained, the share of Uzbekneftegaz JSC is 76.0%, and the sold share is 24.0%. In the draft Decree of the President of the Republic of Uzbekistan "On additional measures to reduce the share of the

Table 2 List of state-owned enterprises whose shares (stakes) in the authorized capital are sold in full to the private sector

No	Name	Territory	State share (%)
1	JSC "Neftegazgazkurilishtamir"	Bukhara Region	36.0
2	JSC "Neftgaztadkikat"	Bukhara Region	36.0
3	JSC "Bukhorogazsanoatkurilish"	Bukhara Region	58.4
4	JSC "Bukhoroneftegazavtonakl"	Bukhara Region	51.0
5	LLC "Alat oil and gas exploration expedition"	Bukhara Region	100
6	JSC "Uztashkineftegaz"	Tashkent	51.0
7	JSC Toshneftegazkurilish	Tashkent	53.3
8	JSC "Uzneftgazinformatika"	Tashkent	52.5
9	JSC "Nefttaminot"	Tashkent	25.0
10	JSC "Sarbon-neftegaz"	Tashkent	51.0

Source Developed by the authors based on [8]

state in the economy", the state share of AndijanNeft JSC is 1.1%, Uzneftgazkazibchikarish JSC—48.3%, Uzneftmahsulot JSC—32.4%, JSC "Uzneftgazishchitaminot"—39.0%, JSC "Uzburguneftgaz"—27.4%, these enterprises are indicated as enterprises liquidated by the state or enterprises with state participation, to which the principle of bankruptcy is applied. To preserve the financial stability of enterprises and ensure the re-employment of employees of the enterprise, we offer management of these enterprises based on public–private partnerships to regional business entities or foreign investors [7].

3 Results

The implementation of the above tasks of the gas transmission system makes it possible to ensure reliable and uninterrupted operation of the industrial part of the facilities, increase export and transit obligations, ensure uninterrupted gas supply to consumers and reduce losses in the process of gas transportation. Global processes in the world economy have led to the fact that the gas industry has become a global industry. The geographical remoteness of deposits from sales markets is approaching with the emergence of new methods of establishing relations between them. Gas markets are poised to grow rapidly. 77.9% of all gas reserves in the world are accounted for by a total of 10 large states (134.0 trillion m^3). By 2020, the volume of gas production in the world will be 5 trillion m^3. Over the next 30 years, global energy demand will increase by 1.8% annually. This is the driving force for the development of demand, which contributes to an increase of 1% of the world's population annually (from the current 6 billion to 8 billion), and the annual economic growth corresponds to 2.5% of GDP (from 2 to 5%, in different ways in different countries). Studies are positive about the prospects for international trade in natural gas. International sales of natural gas by 2030 will amount to about 680–990 billion m^3. Natural gas is a much cleaner energy fuel from an environmental point of view than coal and diesel.

4 Conclusions

The combustion of natural gas does not generate solid pollutants in the form of fly ash, while much less carbon dioxide and sulfur compounds are emitted. This is evidenced by information provided by the Federal Republic of Germany.

Scientific proposals and practical results developed as a result of the research carried out will serve to improve the infrastructure of the industry enterprises and ensure the financial stability of enterprises by reducing the state's share in monopoly sectors implemented in the Republic of Uzbekistan.

References

1. Decree of the President of the Republic of Uzbekistan "On measures to improve the oil and gas sector management system" dated June 30, 2017, Improved the activities of the Joint Stock Company "Uzbekneftegaz".
2. Decree of the President of the Republic of Uzbekistan dated February 7, 2017, No. UP-4947 "On the Strategy of Actions in the Five Priority Areas of Development of the Republic of Uzbekistan in 2017–2021". Appendix 1.
3. Aschauer, D. A. (1989). Is public expenditure productive? *Journal of Monetary Economics, 23*, 177–200.
4. Canning, D., & Pedroni, P. (2004). The effect of infrastructure on long-run economic growth. *The Manchester School, 76*, 504–527.
5. Ishmanova, D. N. (2021). Improving the mechanism of management of infrastructures of enterprises in the oil and gas industry in Uzbekistan. By the example of JSC "Uztransgaz". Doctor of Philosophy thesis. Ziyonetz. Retrieved January 27, 2022 from www.ziyonet.uz.
6. Kucharov, A. S. (2021). New institutions for socio-economic development: The change of paradigm from rationality and stability to responsibility and dynamism. 63–69. https://www.scopus.com/inward/record.uri, https://doi.org/10.1515/9783110699869-007.
7. Oil & Gas. (2021). Retrieved January 27, 2022 from https://www.oilandgasmiddleeast.com/exploration.
8. Statistical data of Uztransgaz JSC of the Republic of Uzbekistan for 2019–2021. Retrieved January 27, 2022 from https://utg.uz/ru/.

Digital Divide and Its Bridging with the Help of GovTech Based on Big Data

Tendencies and Trends in the Process of Digitalization of Personnel Selection by Heads of Commercial Companies

Tatiana E. Lebedeva, Olga V. Golubeva, Zhanna V. Chaykina, Evgeny E. Egorov, and Elena V. Romanovskaya

Abstract The study discusses the essence of digitalization of personnel selection by modern companies in the context of the restrictions associated with the covid-19 pandemic. Today, the development of digital platforms and services in the segment is a topical area, therefore it is recommended to introduce information and digital methods in recruiting. *Purpose*: Development of the implementation and operation of digital services by the heads of commercial organizations ensures the operational personnel management, reduces the costs and the expenses of contacting HR agencies, and also increases the company's competitiveness in the market. The emphasis in the work is made precisely on these aspects of the study. *Methodology*: The study used methods of analyzing the experience of Russian companies in the market in 2020, as well as a survey of the heads of 50 trading companies in Nizhny Novgorod. *Findings*: Based on the conducted two-way analysis, it was decided to use digital services and platforms for personnel selection, however, the respondents insist on a final face-to-face meeting with applicants in order to personally assess their soft-skills, as well as to assess personal qualities. *Originality*: The digitalization of recruiting in the company will allow attracting a larger number of applicants without territorial restrictions and forming effective teams by a professional. The problem of the study is the digitalization of personnel selection processes, but also the caution of company management to use them independently. The authors see the solution to this problem in the constant updating of digital skills.

Keywords Personnel · Digitalization of search · Selection and recruitment of personnel · Pandemic · Digitalization · Recruiting

JEL Codes M5 · J2 · J23 · J24

T. E. Lebedeva (✉) · O. V. Golubeva · Z. V. Chaykina · E. E. Egorov · E. V. Romanovskaya
Minin Nizhny Novgorod State Pedagogical University, Nizhny Novgorod, Russia
e-mail: taty-lebed@mail.ru

E. V. Romanovskaya
e-mail: alenarom@list.ru

V. N. Ostrovskaya and A. V. Bogoviz (eds.), *Big Data in the GovTech System*,
Studies in Big Data 110, https://doi.org/10.1007/978-3-031-04903-3_16

1 Introduction

The pandemic, lockdown and further restrictions associated with covid-19 forced a radical revision of almost all workflows in organizations. New challenges of the time have emerged, information flows have expanded, as life has «shifted to digital». The increase in the speed and digitalization of almost all management processes in commercial organizations is explained by this very circumstance.

The processes associated with the recruitment and selection of personnel also did not stand aside. It should be noted that the current socio-economic characteristics of the personnel market in 2020 has no analogues in history and retrospectives. A particular feature is the fact that commercial companies under coronavirus restrictions were forced to shift to remote working with personnel, including the processes of finding, selecting and hiring them by introducing digital technologies.

In the study, let's consider the experience of large companies in Russia and trading companies in Nizhny Novgorod in the field of digitalization of the processes of search, selection and recruitment of personnel in the era of coronavirus.

As a result, the main tendencies turning into recruitment trends were highlighted.

2 Methodology

A number of researchers define the selection and screening of personnel as two sequential recruitment processes, noting that the screening of personnel is preceded by the selection process, expressed in the measures taken to find and attract candidates [1, 8]. The concept of recruiting is borrowed and identical to the more conservative definition of the recruitment process, however, it is often used as a general term that includes all stages of working with candidates—from searching to making a decision on admission to the vacancy [6, 7].

An analysis of the scientific works of a number of authors has revealed many approaches and grounds for the classification of personnel selection methods.

Recently, the most common classification, which is reflected in the scientific works of a number of scientists [2, 3, 5], is this one, which divides all methods of personnel selection into traditional and non-traditional (Fig. 1).

However, following this classification is not relevant in the modern situation, because now there are many digital platforms that, for example, allow automatic resume creation. Thus, this method moves from the traditional block to the modern one.

The study was carried out on the basis of analysis of secondary information about large companies in Russia and their actions for the selection and recruitment of personnel in the market during the pandemic. Primary information was collected by the survey method, using google forms during March–May 2021. The sample consisted of 50 commercial companies in the city of Nizhny Novgorod. The companies specialize in trading.

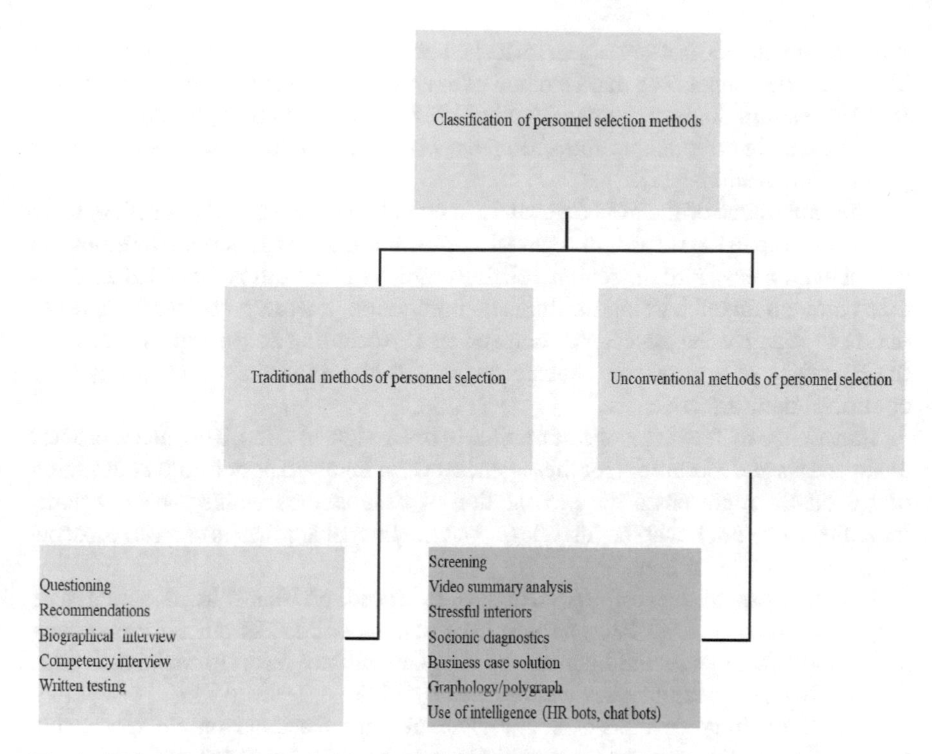

Fig. 1 Classification of personnel selection methods. *Source* Compiled by the authors

3 Results

The first stage of the study was to analyze the experience of large companies, including representative offices of foreign companies in Russia such as Siemens, BASF, Sberbank, Logitech, Technonicol, SAS, Yota during the pandemic in the selection and recruitment of personnel.

For example, Siemens in Russia has translated all interviews with job seekers into online format, previously such interviews accounted for about 12%; has gained experience in recruiting personnel in social networks, whereas they applied only to recruiting agencies until 2020. In addition, the entire document flow was converted to electronic format and an electronic signature was introduced for all employees [9, 10].

The representative of the HR partner of functional divisions and the coordinator of HR projects of the international chemical concern BASF notes that at the very beginning of the pandemic, the company revised its recruitment strategy towards using internal potential and personnel rotation. This was due to the release of the contact area personnel, and employees were given the opportunity to acquire new competencies and try themselves in a different role. Rapid adaptation to changes, as the saying goes, «those who know how to live and work productively in the format of startup companies», is a key aspect in the first place when evaluating applicants when hiring new personnel. The main method of assessing candidates for competencies at BASF is testing. Social networks (Facebook, LinkedIn, Instagram) began to be used for search, as in the example above. Remote work period allowed searching in other geographic locations [11].

The experience of the Yota company seems to be very interesting. In 2020, there was a big jump in the number of regional employees and now almost all divisions can accept employees from the regions. Hiring employees in both 2020 and 2021 takes place through an onboarding meeting and immersion in work processes completely online to identify the necessary competencies. According to the representative of the HR office of the company, such changes will remain in the future as they have confirmed their effectiveness.

Experience of Technonicol corporation: in the middle of 2020, it opened vacancies again and started recruiting, for this, agencies were involved as before. The selection of candidates represented the presentation of case studies online for all regions, including Moscow, but at the final stage—all the finalist applicants came to Moscow for face-to-face interviews.

HR business partner of SAS Russia/CIS faced problems in attracting new employees in early 2020, later, the company opened online HR office for searching and hiring personnel on the Internet, the company did not resort to social networks. The number of new employees was reduced by 22%.

SBS Consulting developed and implemented professional online tests using Google Forms, the proaction.pro platform and launched a new format for interviewing candidates on the VCV video platform, where questions were developed for each position to assess the personal and professional qualities of a candidate, to attract new employees, implemented the Pyrus platform, thanks to which recruiters save data about a candidate—a questionnaire, test results, comments of all interviewers in a single space, to optimize the selection process, this allows for prompt decision-making based on the results of all selection stages and the provision of an offer to the candidate in an online format [11].

Evgeniya Ronzhina, Managing Director, Head of the SberX Headquarters of the Directorate for the Development of the Sberbank Ecosystem, ideologist at Community University, lecturer at the Financial University under the Government of the Russian Federation, noted that in April 2021, together with the Digital Leader networking platform and the CROC company, they conducted a study «Wind of change: New Realities of Digital Interaction» by interviewing more than 500 respondents from 10 industries (IT, industry, banks, public sector, retail, etc.). 56.3% of employees were interviewed online. Half of them felt comfortable communicating online. 45.5% of respondents believe that a recruiter's picture/image quality influences a candidate's opinion about the company. 12% of them are gradually changing their regime and switching to a hybrid format [11].

HR agencies have also transformed their processes in the market. Let's turn to the analysis of their experience. According to the HR Digital 2020 survey, more than a third of companies under the influence of the pandemic began to conduct interviews with job seekers remotely or completely switched to remote recruiting. HeadHunter believes that distance interview format will become mainstream in the near future. Also, according to hh.ru, every fifth company in Russia plans to fully automate the recruitment process within a couple of years. HR specialists use electronic systems both for finding candidates and for conducting interviews and completing test tasks. [10–12].

Poor sound and poor-quality image of the interlocutor are most annoying for survey participants and make it difficult for them to interact during online meetings. These figures very clearly demonstrate modern realities and underline the importance of the right technical equipment.

Therefore, one of the main trends of the 2020s is the automation of routine tasks that recruiters used to do. Viewing resumes, filling out tables, calling candidates and making appointments—today, smart omnichannel platforms based on artificial intelligence are increasingly successfully coping with this. And as the recruitment and initial onboarding process increasingly takes place remotely, these tools work more efficiently, and companies are increasingly investing in them.

Now let's turn to the results of the study of Nizhny Novgorod companies conducted by the authors.

In 2020, 37% of companies did not lose any of their personnel, 18% lost more than 15% of employees during the pandemic, 9%—11–15% of personnel.

At the same time, personnel losses occurred in the segment of marketing departments and sales departments. Additional personnel were required for the contact area of work with consumers, and courier vacancies were also formed. Almost half of the companies, 49%, found themselves in this situation. Companies faced the problems of finding personnel in the context of the fact that recruitment agencies in Nizhny Novgorod were closed at the beginning of the pandemic.

100% of companies resorted to online recruiting: 30% of companies preferred video interviews, 25% of companies offered a practical task and an online survey to candidates for vacant positions, respectively, and 20% offered a solution to ready-made cases (i.e. developed earlier for personnel selection).

The directorate and HR departments of companies are unequivocally inclined towards digitalization of hiring processes and note that the search for optimal methods and platforms has made it possible to find and open a number of services that allow interviewing a large number of candidates in the absence of recruiters [4, 5].

As for online platforms, Skype, Zoom, Teams, other services and platforms that the leaders of organizations were able to master in a short time were actively used when conducting video interviews.

The following advantages were named: interviews with a large number of candidates in a short time, reduced costs for contacting recruiting agencies, reduced time for making managerial decisions. At the same time, many company representatives note that they used the recommendations of friends and searched on social networks when selecting candidates (the most popular was the Instagram network).

The disadvantages of such searches were: low motivation of candidates/or lack of skills for identifying motivation, diagnosing personal qualities of the applicant, lack of psychodiagnostic skills.

85% of company representatives noted that they preferred face-to-face interaction with candidates for the final decision-making on the acceptance of a job seeker, even during such a period.

45% of company representatives came to the conclusion that the pandemic has given a new vector of company renewal and a staff of remote employees, sometimes not even familiar with each other or working in different time zones and can quite effectively solve problems. The main thing is to clearly formulate goals, prescribing the timing of each stage and its budget, coordinate and control, and for companies engaged in trading activities not only in Nizhny Novgorod, but also outside it, this is a chance for growth and survival in difficult conditions.

Most of the executives who participated in the study are extremely wary of expanding the use of digital technology in recruiting. Thus, 53% of the interviewed executives expressed their interest only in using online psychodiagnostics in recruiting, and 28%—in gamified methods of personnel selection, the remaining 19% prefer personal contacts with candidates.

Representatives of Nizhny Novgorod companies associate the main fears of the introduction of digital technologies with the revolution in the use of HR-bots [13, 14], which bring about a decrease in the quality of decisions made; 47% of respondents also have an ambiguous attitude about the use of artificial intelligence in the recruiting system, while 53% noted that that technologies based on artificial intelligence are the future, moreover, they are already being used abroad.

4 Conclusion

Thus, the complete digitalization of the hiring process is controversial and ambiguous in real business. Managers see a risk for the organization with full digitalization, because an employee can pass all the tests and tasks of the system, but will have low labor productivity or not perform his duties at all. At the finish line, applicants must definitely meet with company representatives face-to-face to assess personal qualities and soft-skills—this is their opinion.

Let's highlight a few of the trends that are claiming recruitment trends in the era of covid-19 constraints.

1. Lack of reference to location when searching for candidates, and as a result—expanding the geography of search, a wide pool of candidates, flexible working conditions and no restrictions on relocation.
2. Digitalization of the entire path of the candidate. Much more information about the candidate can be obtained in the same time as before/or build a selection funnel using tools such as testing and video interviews, and so on.
3. Reskilling. Searching and managing talent, as well as moving within the company, it is not enough for an applicant to fit a set of requirements, it is necessary that he had certain skills and shared the company's values, moreover, finding specialists for whom there is no specific position in the staffing table.

Striving for perfect symbiosis: the team and project employees will be able to choose those with whom they are comfortable working. And as a result—a new era in recruitment, where it is necessary to constantly update digital skills, pump more and more changes in the needs and expectations of candidates in the labor market.

References

1. Abolikhina, E. S., & Simanov, M. D. (2018). Principles of digital control in HR. *Postulate, 5–1*(31), 140.
2. Chupina, I. P., Simachkova, N. N., Zarubina, E. V., Fateeva, N. B., & Petrova, L. N. (2020). Digitalization of processes of formation of human resources in HR-management. Moscow Economic Journal. https://cyberleninka.ru/article/n/tsifrovizatsiya-protsessov-formirovaniya-chelovecheskih-resursov-v-hr-menedzhmente/viewer. Accessed September 19, 2021.
3. Dvorskay, E. (2018). Artificial Intelligence in HR: competition with a person or mutually beneficial cooperation. https://vc.ru/future/35297-ii-v-hr-konkurenciya-s-chelovekom-ili-vza imovygodnoe-sotrudnichestvo. Accessed September 19, 2021.
4. Egorov, E. E., Lebedeva, T. E., Prokhorova, M. P., Tsapina, T. N., & Shkunova, A. A. (2020). Opportunities and Prospects of Using Chatbots in HR. Lecture Notes in Networks and Systemsthis link is disabled, 129 LNNS (pp. 782–791).
5. Listikova, A. V., Egorov, E. E., Lebedeva, T. E., Bulganina, S. V., & Prokhorova, M. P. (2020). Research of the best practices of artificial intelligence in the formation and development of personnel. *Lecture Notes in Networks and Systemsthis Link is Disabled* (vol. 73, pp. 1345–1352).
6. Minina, V. N. (2019). HR-bots in human resource management of an organization. *Bulletin of Saint Petersburg University. Management, 3*. https://cyberleninka.ru/article/n/hr-boty-v-upravl enii-chelovecheskimi-resursami-organizatsii/viewer. Accessed September 19, 2021.
7. Nikonorova, A. V., & Toropova, N. N. (2018). Problems and methods of increasing efficiency in the implementation of the personnel selection process. *Bulletin of the Russian State University for the Humanities. Series Economy. Control. Right, 3*(13), 90–102.
8. Prokhorova, M. P., Lebedeva, T. E., Ksenofontova, A. I., & Bobarykin, D. A. (2020). Methods for evaluating soft skills when selecting personnel. *Moscow Economic Journal, 4*, 49.
9. Romanovskaya, E. V., Andryashina, N. S., Kuznetsova, S. N., Smirnova, Z. V., & Ivonina, O. G. (2021). Digital technologies in Russia: Trends, place and role in economy. *Lecture Notes in Networks and Systems, 155*, 344–351.
10. Saifullina, L. D., & Komnatnaya, A. V. (2020). The impact of the pandemic on the automation of the main tasks of an HR specialist. *Economics and Business: Theory and Practice*, 9–2. https://cyberleninka.ru/article/n/vliyanie-pandemii-na-avtomatizatsiyu-osn ovnyh-zadach-hr-spetsialista/viewer. Accessed September 19, 2021.
11. Semina, A. P. (2020). Digitalization of personnel management processes: SMM in HR. *Discussion, 1*(98). https://cyberleninka.ru/article/n/tsifrovizatsiya-protsessov-upravleniya-per sonalom-smm-v-hr/viewer. Accessed September 19, 2021.
12. Smirnova, Zh. V., & Kochnova, K. A. (2019). Training of employees of service enterprises using information technologies. *Vestnik of Minin University, 7, 1*(26), 5.
13. Vakulenko, R. Y., Tyumina, N. S., Potapova, E. A., & Proskulikova, N. (2016). Analysis of organizational and technological environment of the existence of electronic services. *Vestnik of Minin University, 1–1*, 1.
14. Yashkova, E. V., Sineva, N. L., Semenov, S. V., Kuryleva, O. I., & Egorova, A. O. (2020). The impact of digital technologies on various activity spheres and social development. *Lecture Notes in Networks and Systems, 91*, 149–155.

Long-Term Effects of COVID-19: How the Pandemic Highlighted the Global Digital Divide

Elena E. Kukina, Natalia M. Fomenko, Olga F. Alekhina, Elena V. Smirnova, and Olga A. Pecherskaya

Abstract The article is a study of the prerequisites for the formation of the global digital divide, as well as its structural features, including during the COVID-19 pandemic. The authors identify technological, economic and social components of the global digital divide. The authors analyze the indicators assessing the technological component of the global digital divide for the period 2015–2020. The authors used indicators of the dynamics of the global ICT market and the dynamics of the e-commerce market to study the economic component of the digital divide and the impact of the pandemic on it. It is concluded that the pandemic has increased thc importance of digital technologies in general, as well as opportunities to reduce the global digital divide. The social component of the digital divide is manifested in the different levels of realization of the opportunities of individuals and households with the help of ICT technologies. The authors proved that the gap in access to digital technologies and their use contributes to the emergence of a divide in knowledge and human development opportunities. The article proposes public policy measures aimed at bridging the global digital divide.

Keywords Digitalization · Information and communication technologies · Global digital divide · Lag · Pandemic · Human capital · Opportunities

---

E. E. Kukina (✉)
The Financial University under the Government of the Russian Federation (Lipetsk Branch), Lipetsk, Russia
e-mail: economresearch@mail.ru

N. M. Fomenko
Plekhanov Russian University of Economics, Moscow, Russia

O. F. Alekhina
National Research Lobachevsky State University of Nizhny Novgorod, Nizhny Novgorod, Russia

E. V. Smirnova
Kuban State Technological University, Krasnodar, Russia

O. A. Pecherskaya
Voronezh State University of Forestry and Technologies named after G.F. Morozov, Voronezh, Russia

V. N. Ostrovskaya and A. V. Bogoviz (eds.), *Big Data in the GovTech System*, Studies in Big Data 110, https://doi.org/10.1007/978-3-031-04903-3_17

JEL Codes F01 · O33

1 Introduction

Intensive digitalization has provided multiple advantages to modern society. In particular, developed countries have the opportunity to consolidate their already strong positions in the global space. Outsider countries have found new digital tools to overcome the lag in their development, bypassing several intermediate stages at once [21, 25]. However, access to the Internet, available digital technologies and the ability of economic entities to use them to realize their interests have become factors determining the widening of the development gap at the country and global levels. According to world practise, overcoming the global digital divide depends on the effectiveness and efficiency of the measures taken within the framework of the implemented state policy, as well as the purposeful cooperation of all interested parties—authorities, international organizations, civil society institutions, the public and the private sector of the economy [20]. However, currently, a new external circumstance has appeared—a pandemic, it has radically changed the balance of forces in the global space. The COVID-19 pandemic has become both a threat to the world community and a trigger for the transition to a new qualitative level of digitalization for many countries. The countries that have intensified the efforts of the state and society towards the use of digital technologies under quarantine restrictions have been able to make a digital breakthrough [23, 24]. Other countries that have remained aloof from new trends and technologies have moved to the stage of a deep post-pandemic crisis. Thus, the distribution of forces in the global economy has changed during the pandemic. Probably, in such conditions, the global digital divide has undergone significant transformations.

2 Literature Review

Currently, modern societies, along with economic, social and cultural inequality, experience inequality associated with the uneven distribution of information and modern information and communication technologies [25]. In this case, we are talking about the digital divide as the gap between those who have regular and effective access to modern information and communication resources, and those who do not have free, unhindered access to information resources for various reasons [19]. The most common reason is government policies that do not support digital development goals. This, in turn, manifests itself in the lack of state support for projects to improve the digital infrastructure, and, as a result, its uneven development [3]. Some researchers point to the social reasons for the deepening of the digital divide associated with social stratification, manifested in unequal access to the Internet and its use [27]. Accelerating the pace of digitalization exacerbates the problem of digital inequality. Digital technologies have already led to the creation of noticeable wealth in a very short time, and digital dividends are concentrated in a small number of countries, companies, individuals [32]. However, new non-standard circumstances related to the COVID-19 pandemic, lockdown, quarantine restrictions and social distancing introduced in most countries have pushed people to the widespread use of digital technologies. This provoked the intensive development of digital platforms, as well as online services in various spheres of human life [22].

3 Methodology

The purpose of the study is to identify the prerequisites for the formation of the global digital divide, as well as to establish the role of the COVID-19 pandemic as a factor determining its transformation. Research objectives: (1) to investigate the content of the concept of "global digital divide" and to reveal the methodological features of its definition; (2) assess the impact of the pandemic on the existing global digital divide; (3) propose public policy measures aimed at bridging the digital divide. Research methods: theoretical analysis, systematization method, comparative analysis, economic and statistical analysis, graphical method, systematic approach.

4 Results

There are three key components of the digital divide: (1) the technological component of the digital divide is associated with the level of digital infrastructure development, the presence/absence of networks, access to the Internet, mobile communications, the level of automation and virtualization of activities; (2) the economic component of the digital divide is due to the development of the ICT sector, the volume and dynamics of investments in the development of the digital sector, and the growth rate of the digital economy. At the enterprise level, the prerequisites for the digital divide may be income, the availability of ICT skills and competencies of specialists, the development of digital ecosystems in business; (3) the social component of the digital divide is determined by prerequisites related to age, gender, education, and social status [32]. Accordingly, this creates new forms of poverty and exclusion, as well as reproduces existing inequalities and social differences [7]. Separate indicators of international indices and ratings can be used to characterize the various components of the global digital divide (Table 1).

The table shows that in general, in the period 2015–2020, there is a positive trend in improving digital infrastructure and expanding access to the Internet for various categories of users in all countries. The strongest digital gap between developed and developing countries is observed in terms of fixed-telephone subscriptions (4.6 times in 2019), mobile broadband (1.9 times in 2020), fixed broadband (2.9 times in 2020), households with Internet access at home (1.8 times in 2019). However, the

Table 1 Some indicators of the technological component of the global digital divide, 2015–2020

Region	2015	2016	2017	2018	2019	2020
Fixed-telephone subscriptions, per 100 inhabitants						
World	14.0	13.4	12.9	12.4	11.9	–
Developed	39.0	38.0	37.0	35.6	34.3	–
Developing	8.9	8.3	8.0	7.7	7.4	–
Least developed countries	0.9	0.9	0.8	0.9	0.8	–
Mobile-cellular telephone subscriptions, per 100 inhabitants						
World	97.3	100.6	102.7	104.9	107.8	105.0
Developed	124.5	125.9	126.3	128.5	131.8	133.4
Developing	91.6	95.4	97.9	100.1	103.0	99.3
Least developed countries	67.5	67.6	68.4	71.4	74.9	74.0
Active mobile-broadband subscriptions						
World	44.6	51.9	62.8	69.5	74.2	75.0
Developed	89.2	97.0	108.7	116.6	123.9	125.2
Developing	35.4	42.7	53.5	60.1	64.3	65.1
Least developed countries	14.9	19.9	26.2	29.0	31.8	33.2
Fixed-broadband subscriptions						
World	11.4	12.2	13.6	14.0	14.8	15.2
Developed	29.5	30.4	31.5	32.2	33.2	33.6
Developing	7.6	8.5	9.9	10.3	11.1	11.5
Least developed countries	0.8	0.9	1.0	1.1	1.2	1.3
Population covered by a mobile-cellular network						
World	94.8	95.3	96.1	96.3	96.7	96.7
Developed	98.5	98.6	98.7	98.7	99.6	99.6
Developing	94.0	94.6	95.5	95.8	96.2	96.1
Least developed countries	86.1	87.1	87.1	87.9	88.4	88.9
Households with Internet access at home						
World	47.9	50.5	53.4	55.7	57.4	–
Developed	80.1	81.3	83.1	84.5	85.2	–
Developing	36.5	39.7	43.0	45.8	47.8	–
Least developed countries	9.7	11.5	13.6	15.1	16.3	–

Source Compiled by the authors based on [17]

digital gap between developed and developing countries is practically not noticeable in terms of such indicators as "mobile-cellular telephone subscriptions", "population covered by a mobile-cellular". The least developed countries have low rates of Internet penetration and network bandwidth, and access by the population and households is very limited. Among the reasons for this lag, experts note, first of all, the high cost of new technologies and means of communication, as well as the lack of a technological base that contributes to the transformation of existing and the formation of new knowledge, skills and competencies necessary for the widespread use and application of modern technologies [16, 25].

Let's randomly select several countries of the world, for example, China, the USA, Germany, France, Sweden, Estonia, Russia, Turkey, Indonesia, and South Africa to illustrate the technological component of the digital divide. According to the ICT Development Index, Sweden is among the most "connected" countries, characterized by a high level of penetration of fixed and mobile communication services, in the country the majority of the population uses the Internet. During the pandemic, the Government continues to focus its efforts on the deployment of broadband communications and ensuring conditions for the functioning of a competitive and modern telecommunications market [15]. China is the world's largest telecommunications market in terms of the number of mobiles, fixed telephone, fixed broadband and mobile broadband contracts; it is also a leading exporter of ICT products. The leadership role of the Government, the activities of the private sector and the extensive production base of ICT have combined to effectively eliminate the digital divide between urban and rural areas of the country and the successful development of Internet applications, such as e-commerce in rural areas. Telecommunications in the USA is a stable sector, all its segments are characterized by high competition, and penetration levels are among the highest among all types of services. However, there remain obstacles that need to be overcome, in particular, differences in access to broadband between urban areas and more remote areas of the country. This created certain difficulties during the lockdown period of 2020. However, investments and infrastructure development in the pandemic continued, taking into account the desire of operators to use new telecommunication technologies and provide better service. Germany is among the world leaders in the field of ICT, having a well-developed ICT infrastructure and a high level of ICT penetration into households. The efforts made by the Estonian Government have turned the country into one of the most "connected" countries in Europe and the world. Internet usage rates and the degree to which households are provided with ICT services are extremely high. In 2020, the development of ICT continued: the government sought to provide the best conditions for private operators to invest in a new generation of networks. The Russian Federation has a dynamically developing telecommunications market [3]. The regulatory body seeks to bridge the digital divide between regions and provide the population with modern telecommunication services by promoting infrastructure modernization. Nevertheless, according to some indicators of ICT development, the country is noticeably significantly behind the leaders. Turkey has a relatively large telecommunications market with huge growth potential. Mobile and fixed-line penetration rates are below the European average, however, are growing rapidly. South Africa is at the forefront of the technological development of the region, with the latest broadband technologies and wide coverage. This is facilitated by an appropriate regulatory framework and a competitive market with the participation of the private sector. In 2020, the South African government unveiled a ten-year investment plan, including in the development of digital infrastructure, amounting to $133 billion to overcome the pandemic crisis [12]. Some time ago, the competition was introduced in Indonesia, which led to the wider use of mobile and broadband services.

Various indicators can be used to describe the economic component of the digital divide. For example, the volume of final products produced in the ICT sector, the volume of investments in the development of digital technologies, the dynamics of the e-commerce market. The most significant contribution to the development of information and communication technologies in 2018 was made by the United States ($1.3 trillion in ICT costs), China ($499 billion). In addition, Japan, Great Britain and Germany are among the top five, but the Philippines (+7.5%), India (+7%) and Peru (+6.7%) demonstrate the highest CAGR indicators [2, 34]. The pandemic has radically changed the situation on the world market. The restrictions imposed in the countries caused a decrease in demand for goods and services from both the population and organizations [14]. Experts attributed the decline of the ICT industry to the COVID-19 pandemic, which caused companies to cut budgets for equipment, software and services. It was the large-scale spread of coronavirus infection that contributed to digitalization: businesses had to transfer staff to remote work, and educational institutions to distance learning, which spurred demand for cloud and other IT tools. In 2019, the unequivocal leader in the e-commerce market was China (1,934.8 billion US dollars), which has more than a three-fold margin compared to the USA following it (586.9 billion US dollars) [8, 13]. Thus, before the pandemic, the digital divide between developed and developing countries was obvious. The coronavirus pandemic has had a significant impact on e-commerce and consumer behaviour on the Internet around the world [8]. In the pandemic 2020, the e-commerce market in the world grew by 18% compared to 2019 [39]. The pandemic highlighted the importance of digital technologies in general, as well as opportunities for expanding international cooperation to facilitate the cross-border movement of goods and services, reduce the digital divide. Let's turn to the Networked Readiness Index (NRI) to illustrate the economic component of the global digital divide. The top 10 countries in the NRI in pandemic 2020 (Sweden, Denmark, Singapore, Netherlands, Switzerland, Finland, Norway, the USA, Germany, and the UK) were also in the top 10 in 2019. Some regions continue to lag. Most notably, Africa lags behind all regions, especially concerning the use of ICT. As soon as the COVID "ripple effect" begins to affect international trade and investment flows, such discrepancies between "networked economies" and "laggards" may be intensified [26]. Sweden (82.75 scores), unlike most developed European countries, has chosen a different strategy of behaviour in the face of a pandemic. The country abandoned quarantine restrictions, which allowed it to prevent a deep economic crisis and continue implementing an effective system of working with personnel and conducting scientific research. In 2020, Sweden ranked second in the Global Innovation Index (GI) ranking [38]. The largest Swedish companies in the digital sector—Ericsson, Hexagon, Telia Company—strengthened their positions during the pandemic, focusing their activities on maintaining the well-being of the population [5, 18]. The latest technologies and digitalization of business have played an important role in reducing the spread and mitigating the impact of the pandemic on society and business in China (58.44 scores) [29]. The active use of LBS services, big data analysis and robotics has made it possible to track and identify high-risk cases, restrict the movement of people and minimize between people, limiting the territory of the spread of the virus. Chinese companies everywhere used digital technologies to resume operations, live broadcasts of events supported the interest of consumers, IoT and robotics allowed to accelerate the automation of production, digital technologies were indispensable for the organization of remote work of employees [11]. In the United States (78.91 scores), the pandemic period was marked by the establishment of Government partnerships with digital platforms IBM, Google, Amazon and Microsoft to allow researchers to perform a large number of calculations in the field of epidemiology, bioinformatics and molecular modelling. In addition, stimulus measures for economic recovery are being implemented through a series of aid packages. About $200 billion has been allocated for the development of telemedicine [11]. Digital technologies have formed the basis of a wide range of tools to combat the spread of the pandemic and its economic consequences in Germany (77.48 scores). In 2020, the country adopted a program of economic assistance to overcome the crisis after the COVID-19 pandemic of 50 billion euros. The key areas of support for digital projects are security (10 billion euros), healthcare, including epidemic response systems, telemedicine and medical robotics (7 billion euros), distance learning and reducing digital inequality (4 billion euros). The fight against the epidemic in Estonia (70.32 scores) is carried out based on such methods as quarantine, testing, contact tracking and the efforts of doctors using IT industry technologies. The impeccable work of the E-government and the availability of data on citizens helps to maintain the efficiency of the administrative apparatus and contain the spread of coronavirus infection in the country [10]. In Russia (45.23 scores), a system of countering the pandemic has been built, including using digital technologies. On the one hand, the crisis has shown the maturity of the Russian information technology and infrastructure industry and the ability to mobilize the necessary resources. On the other hand, the pandemic has had a negative impact on the IT industry in the form of a decrease in the volume of the industry's products sold. The package of measures to support the IT industry proposed by the Government of the Russian Federation involves reducing the insurance premium rate for IT companies from 14 to 7.6% and the income tax rate from 20 to 3%. The Turkish authorities administratively and financially supported the aspirations of Turkish companies for digital transformation to prepare the country's economy for further digitalization in the face of a pandemic [30]. During the epidemic, online purchases of goods of different categories increased by 10–34%, and sales of branded goods in the first days of the curfew increased by 14% at once. At the same time, 37.4% of buyers tried online shopping for the first time [35]. Representatives of the Turkish tourism sector have launched a project to create a single online platform Goturkey.com, which presents the cultural and tourist attractions of the country in an online format [31]. Digital technologies have also been used to improve the effectiveness of sanitary measures [28]. The pandemic and self-isolation have triggered the growth of e-commerce in South Africa (45.26 scores). Sellers began to adapt and shift the focus towards online sales, and brands initiated communication with the audience using information mechanisms, introduction to the online, adaptation of content in social networks [37]. The coronavirus pandemic has accelerated the digitalization of business in Indonesia (46.71 scores). As a result, the online sales market began to grow rapidly. A struggle for a share in the Indonesian market has unfolded between representatives of the e-commerce segment and FinTech startups [1]. According to IPrice and App Annie, in 2020, changes in the choice of contactless payment market participants during the pandemic led to an increase in the number of users of financial applications by 70%.

The social component of the digital divide manifests itself in different levels of realization of the opportunities of individuals and households with the help of digital technologies. Stiakakis et al. [33] emphasized that the gap in access to and use of digital technologies contributes to the emergence of a gap in knowledge and development opportunities. Thus, using the advantages of the digital environment and Internet resources allows individuals to gain knowledge, education and realize their potential. In turn, the new quality of human capital can provide more income and increase productivity [9, 21] (Table 2).

In developed countries, the values of Internet usage indicators by individuals are significantly higher. Meanwhile, in the period 2015–2019, there is a positive dynamics of growth of this indicator in all countries. The use of the Internet contributes to the growth of e-government indicators, that is, the expansion of the population's ability to use digital platforms to make payments and receive public services. The values of this indicator increased in all countries in 2016–2019. An important indicator reflecting the impact of digital technologies on human well-being is the Human Development Index.

The solution to the problem of strengthening the global gap is seen in the implementation of differentiated policies for two groups of countries, depending on the current state of digitalization and its speed: (1) For leading countries in terms of digitalization (Sweden, China, USA, Germany, Estonia): support of enterprises' initiatives to introduce digital tools for attracting consumers (e-commerce, digital payments, entertainment); training and retention of IT personnel; stimulating the creation of digital startups; providing fast and public Internet access; specialization in the export of digital goods and services; coordination of the activities of participants in the innovation process; investing in the development of the institutional environment and the regulation of capital markets to support innovation; formation of consumer protection tools against privacy violations, cyber-attacks; improving immigration policy; the development of ecosystems that generate innovation in these areas. (2) For lagging countries in terms of digitalization (Russia, Indonesia, Turkey, and South Africa): improving mobile Internet access; strengthening the institutional environment and development of digital legislation; encouraging investment in digital enterprises, financing digital R&D, training IT personnel; measures to reduce inequalities in access to digital tools by gender, class, ethnicity and geography; long-term investments in solving basic infrastructure problems; creating of applications to solve urgent problems of the population, the spread of digital tools (digital payment platforms) [6].

Table 2 Some indicators of the social component of the digital divide, 2015–2020

Region	2015	2016	2017	2018	2019	2020
Individuals using the Internet (%)						
World	41.1	43.9	46.3	49.0	51.4	–
Developed	76.8	81.0	81.9	84.9	86.7	–
Developing	33.7	36.3	39.1	41.9	44.4	–
Least developed countries	12.4	14.3	16.1	17.6	19.5	–
Human Development Index						
Sweden	0.913	–	0.933	0.937	0.945	–
China	0.738	–	0.752	0.758	0.761	–
USA	0.920	–	0.924	0.920	0.926	–
Germany	0.926	–	0.936	0.939	0.947	–
Estonia	0.865	–	0.871	0.882	0.892	–
Russia	0.804	–	0.816	0.824	0.824	–
Turkey	0.767	–	0.791	0.806	0.820	–
Indonesia	0.689	–	0.694	0.707	0.718	–
South Africa	0.666	–	0.699	0.705	0.709	–
E-Government Index						
Sweden	–	0.8704	–	0.8882	0.9365	–
China	–	0.6071	–	0.6811	0.7948	–
USA	–	0.8420	–	0.8769	0.9471	–
Germany	–	0.8210	–	0.8765	0.8524	–
Estonia	–	0.8334	–	0.8486	0.9473	–
Russia	–	0.7215	–	0.7969	0.8244	–
Turkey	–	0.5900	–	0.7112	0.7718	–
Indonesia	–	0.4478	–	0.5258	0.6612	–
South Africa	–	0.5546	–	0.6618	0.6891	–

Source Compiled by the authors based on [4, 36]

5 Conclusion

Firstly, the authors clarified the content of the concept of the "global digital divide", highlighting technological, economic and social components. Secondly, the analysis of the dynamics of indicators was carried out, which allowed the authors to assess the technological, economic and social components of the global digital divide for the period 2015–2020 in groups of countries (developed, developing, least developed), as well as on the example of countries of different levels of development. It was revealed that, on the one hand, the pandemic contributed to accelerated digital transformations, countries began to actively switch to the use of information technologies and new opportunities appeared to accelerate digital development. On the other hand, with the simultaneous use of the potential of digital technologies by all countries, the existing digital divide may persist. Thirdly, the authors proposed a matrix of differentiation of public policy measures to contain the global digital divide during the pandemic.

Data Availability

1. Data on indicators of the technological component of the global digital divide, 2015–2020 are available in https://figshare.com/, https://doi.org/10.6084/m9.figshare.16909018.
2. Data on indicators of the social component of the global digital divide, 2015–2020 are available in https://figshare.com/, https://doi.org/10.6084/m9.figshare.16909021.

References

1. Bloomchain. (2020). Indonesia's digitalization has provoked a struggle for the market among local companies. https://bloomchain.ru/newsfeed/tsifrovizatsija-indonezii-sprovotsirovala-borbu-za-rynok-sredi-mestnyh-kompanii. Accessed September 25, 2021.
2. Bsc-consulting. (2018). In Global ICT spending will approach $4 trillion. https://bsc-consulting.ru/blog/analytics/050218/. Accessed September 25, 2021.
3. Bychkova, N., Tavbulatova, Z., Ruzhanskaya, N., Tamov, R., & Karpunina, E. (2020). Digital readiness of Russian regions. In *Proceeding of the 36th IBIMA Conference* (pp. 2442–2461). Granada, Spain.
4. Center for Humanitarian Technologies. (2021). Human Development Index 2006–2021. https://gtmarket.ru/ratings/human-development-index. Accessed September 25, 2021.
5. Demidova, E. (2018). Features of digitalization of the Scandinavian region. file:///C:/Users/Downloads/osobennosti-tsifrovizatsii-stran-skandinavskogo-regiona.pdf. Accessed September 25, 2021.
6. Digital Evolution Scorecard. (2021). Digital Intelligence Index. https://digitalintelligence.fletcher.tufts.edu/trajectory. Accessed September 25, 2021.
7. DiMaggio, P., Hargittai, E., Celeste, C., & Shafer, S. (2004). Digital inequality: From unequal access to differentiated use. In: K. Neckerman (Ed.), *Social inequality*. Russell Sage Foundation.
8. Emarketer. (2019). Global Ecommerce 2019. https://www.emarketer.com/content/global-ecommerce-2019. Accessed September 09, 2021.
9. Hargittai, E. (2002). Second level digital divide: Differences in people's online skills. *First Monday, 7*(4). http://www.eszter.com/research/pubs/hargittai-secondleveldd.pdf. Accessed September 25, 2021.
10. Hightech. (2020). How boring technologies help Estonia in the fight against the pandemic. https://hightech.plus/2020/06/21/kak-skuchnie-tehnologii-pomogayut-estonii-v-borbe-s-pandemiei. Accessed September 25, 2021.
11. HSE. (2020). Digest of foreign practices on economic recovery from the crisis against the background of a pandemic: new technological policy. https://www.hse.ru/mirror/pubs/share/384494070.pdf. Accessed September 25, 2021.
12. Interfax. (2020). South Africa unveils $133 billion infrastructure investment plan. https://www.interfax.ru/business/714454. Accessed September 25, 2021.
13. IPGResearch. (2019). Global development of e-commerce. https://rgud.ru/documents/2020-IPG.Research_E-commerce.pdf. Accessed September 09, 2021.
14. IssekHSE (2021). The ICT sector has developed immunity to COVID overloads. https://issek.hse.ru/news/446639217.html. Accessed September 09, 2021.

15. ITU. (2018). Report Measuring the Information Society. A brief overview. 2018. https://www.itu.int/en/ITU-D/Statistics/Documents/publications/misr2018/MISR_Vol_2_R.pdf. Accessed September 25, 2021.
16. ITU. (2020). Household Internet access in urban areas is twice as high as in rural areas. https://www.itu.int/en/mediacentre/pages/pr27-2020-facts-figures-urban-areas-higher-internet-access-than-rural.aspx. Accessed September 25, 2021.
17. ITU. (2021). Statistics. https://www.itu.int/en/ITU-D/Statistics/Pages/stat/default.aspx. Accessed September 25, 2021.
18. Karpunina, E., Beilina, A., Butova, L., Trufanova, S., & Astakhin, A. (2020). Towards Sustainable Development through Bridging Digital Penetration Gaps. In *Scientific and Technical Revolution: Yesterday, Today and Tomorrow. Lecture Notes in Networks and Systems* (pp. 476–486). Springer.
19. Karpunina, E., Agabekyan, R., Petrov, I., Gorlova, E., & Sobolevskaya, T. (2022). BRICS countries as new growth poles of the global digital economy. *International Journal of Economic Policy in Emerging Economies*. https://doi.org/10.1504/IJEPEE.2021.10036746.
20. Karpunina, E., Gubernatorova, N., Daudova, A., Stash, Z., & Kargina, L. (2020). the spillover effects of the digital economy. In *Proceedings of the 36th IBIMA Conference* (pp. 942–954). Granada, Spain.
21. Karpunina, E., Konovalova, M., Titova, E., Kheyfits, B., & Sobolevskaya, T. (2020). New paradigm of the strategy of advanced development in the digital economy: Prerequisites, contradictions and prospects. In: *Proceeding of the 35th IBIMA conference* (pp. 2270–2282). Seville, Spain.
22. Karpunina, E., Magomaeva, L., Kochyan, G., Ponomarev, S., & Borshchevskaya, E. (2021). Digital inequality and forms of its appearance: A comparative analysis in the OECD and BRICS countries. In *Proceeding of the 37th IBIMA Conference* (pp. 1028–1040), Cordoba, Spain.
23. Karpunina, E., Moskovtceva, L., Zabelina, O., Zubareva, N., & Tsykora, A. (2021). *Socio-economic impact of the covid-19 pandemic on OECD countries*. Research in Economic Anthropology, Emerald Publishing Limited.
24. Mejokh, Z., Korolyuk, E., Sozaeva, D., Pilipchuk, N., & Karpunina, E. (2020). Economic security of Russian regions: Risk factors and consequences of the covid-19 pandemic. In *Proceeding of the 36th IBIMA Conference* (pp.8197–8205). Granada, Spain.
25. Perfilieva, O. (2007). The problem of the digital divide and international initiatives to overcome it. *Bulletin of International Organizations, 2*(10), 34–48.
26. Portulance Institute. (2021). The network readiness index 2020. Accelerating digital transformation in a post-COVID global economy. https://networkreadinessindex.org/wp-content/uploads/2020/11/NRI-2020-V8_28-11-2020.pdf. Accessed September 25 2021.
27. Ragnedda, M., & Muschert, G. (2013). *The digital divide: The Internet and social inequality in international perspective*. Routledge.
28. RIA. (2020). Turkey has introduced a digital code for travelling in transport and staying in a hotel. https://ria.ru/20200930/turtsiya-1578001681.html. Accessed September 25, 2021.
29. Roscongress. (2020). China's Experience: digital technologies at the forefront of the fight against COVID-19. https://roscongress.org/materials/opyt-kitaya-tsifrovye-tekhnologii-na-peredovoy-borby-s-covid19/. Accessed September 25, 2021.
30. Rossaprimavera. (2020). The Turkish authorities decided to help the digital transformation of business in the country. https://rossaprimavera.ru/news/5529299c. Accessed September 25, 2021.
31. Rusturkey. (2020). Modern technologies and digitalization will increase the tourist flow to Turkey. https://rusturkey.com/post/210806/sovremennye-tehnologii-i-cifrovizaciya-povysyat-turpotok-v-turciyu. Accessed September 25, 2021.
32. Safiullin, A., & Moiseeva, O. (2019). Digital Inequality: Russia and other countries in the Fourth industrial revolution. *St Petersburg State Polytechnical University Journal Economics, 12*(6), 26–37.
33. Stiakakis, E., Kariotellis, P., & Vlachopoulou, M. (2010). From the digital divide to digital inequality. A secondary research in the European union: e-Democracy 2009. *LNICST, 26*, 43–54.

34. Tadviser. (2021). Gartner: IT market will sink by 8% due to COVID-19. https://www.tadviser.ru/index.php/Статья:ИТ_(мировой_рынок). Accessed, September 25, 2021.
35. Turkestate. (2020). There is a sharp increase in online trading in Turkey. https://turk.estate/news/v-turtcii-nablyudaetsya-rezkij-rost-onlajn-torgovli. Accessed September 25, 2021.
36. United Nations. (2020). E-Government Survey 2020. Digital Government in the Decade of Action for Sustainable Development. https://publicadministration.un.org/egovkb/Portals/egovkb/Documents/un/2020-Survey/2020%20UN%20E-Government%20Survey%20(Full%20Report).pdf. Accessed September 25, 2021.
37. Volyansky, A. (2020). How the coronavirus and the crisis will affect the African economy. https://vc.ru/u/500138-anton-volyanskiy/123335-kak-koronavirus-i-krizis-povliyayut-na-ekonomiku-afriki. Accessed September 25, 2021.
38. WIPO. (2020). Publication of the 2020 Global Innovation Index. https://www.wipo.int/global_innovation_index/ru/2020/. Accessed September 25, 2021.
39. World Trade Organization. (2020). E-commerce, trade and the Covid-19 pandemic. https://www.wto.org/english/tratop_e/covid19_e/ecommerce_report_e.pdf. Accessed September 25, 2021.

Youth, Work and Skills: A Map of the Transformation of the Workforce and Employment

Asya E. Arutyunova, Rustam I. Khanseviarov, Elena A. Okunkova, Patimat R. Alieva, and Elena A. Makareva

Abstract The research aims at identifying the features of the transformation of the labour market in terms of employment of modern youth, as well as determining the vector of development of professional skills of young people to ensure their sustainability in the labour market during the pandemic. The article confirms three hypotheses: (1) the labour market is changing, the main trends in the labour market are currently due to digitalization and the impact of the pandemic; (2) youth are more vulnerable to the effects of the pandemic, but adapt faster to changes; (3) the use of the labour force transformation map and employment becomes a prerequisite for the development of tools to support and adapt youth during the pandemic. The employment of young people on a global scale and by groups of countries is analyzed. The reduction of employed youth by 27.8% in the period 1994–2021 was revealed, its causes were revealed. A map of the transformation of the labour force and employment is presented. It can be used to determine the way of professional development of modern youth and to determine state support measures. A comprehensive toolkit is proposed for the adaptation of young people to the labour market in terms of a pandemic based on a map of the transformation of the labour market and employment.

A. E. Arutyunova (✉)
Kuban State Technological University, Krasnodar, Russia
e-mail: aru-asya@yandex.ru

R. I. Khanseviarov
Samara State University of Economics, Samara, Russia
e-mail: rust1978@mail.ru

E. A. Okunkova
Plekhanov Russian University of Economics, Moscow, Russia

P. R. Alieva
Dagestan State University, Makhachkala, Russia

E. A. Makareva
Voronezh State University of Forestry and Technologies Named After G.F. Morozov, Voronezh, Russia

V. N. Ostrovskaya and A. V. Bogoviz (eds.), *Big Data in the GovTech System*, Studies in Big Data 110, https://doi.org/10.1007/978-3-031-04903-3_18

Keywords Youth · Pandemic · Competencies · Skills · Digitalization · Workforce · Adaptation

JEL Codes J13 · J21

1 Introduction

The COVID-19 pandemic has shaken the global economy and dramatically changed the situation in the labour market [17, 18]. The adaptation tools to it were the reduction of employees, then transfer to a remote work format, mechanisms of part-time employment, as well as changes in the conditions of remuneration [13, 34]. The pandemic has actualized the issues of implementing the personnel policy of organizations: factors of labour productivity and quality of work have come to the fore. In conditions when organizations have to give up part of the staff to survive in a crisis, the choice is made in favour of more experienced, professionally competent, stress-resistant and adaptive employees.

The impact of the pandemic affects population groups and categories of workers in different ways. In particular, young people were in the most vulnerable position. This is due to the increasing problems related to the realization of their right to employment, access to the labour market in terms of reduced production and mass bankruptcy of enterprises, as well as lack of work experience and lack of professional competencies [12]. According to ILO research conducted in 2020, one in five young people in the world lost their jobs [29]. In addition, young people are faced with the problem of reducing wages and strengthening employers' requirements for additional training to acquire the necessary skills and abilities.

At the same time, young people have absolute advantages over other categories of the population, which will be able to provide them with competitiveness and stability in the labour market. This is high adaptability to changing conditions, a tendency to continuous self-education, as well as the possession of digital skills [30]. Thus, taking into account the factors of labour force transformation and employment will allow young people to overcome the challenges of the pandemic crisis and digitalization in a short time.

2 Literature Review

The key characteristics of youth as a special category of the population with a certain type of behaviour in the labour market are presented in the works of [10, 20, 23, 25, 41]. In particular, the researchers note that young people are an age group of people in the range of 16–25 years, who are characterized by a transitional position and a high level of mobility and adaptability, a low level of personal well-being and low social status, as well as an active search for their place in life and profession.

The causes and problems associated with the entry of young people into the labour market and the development of youth unemployment are reflected in the works of [7, 26, 40].

Digitalization is one of the key modern trends in the transformation of the labour market and employment. It changes the requirements for the competencies and skills of specialists and threatens those employees who do not meet the new qualities. In particular, according to the new labour and employment model "Work 4.0", there is a demand for labour with high qualification requirements, the so-called "technological change based on skills" [26, 32].

Changes in the labour market under the conditions of digitalization are disclosed in the works of [3, 6].

A new stage of labour market transformations and the emergence of employment problems of modern youth is associated with the impact of the pandemic. In these conditions, there is a blurring of the classical labour organization and the usual employment schemes, as well as distortions in employment and unequal attitude of employers to different age groups [22, 26].

3 Methodology

The purpose of the study is to identify the features of the labour market transformation in terms of modern youth employment, as well as to determine the way of development of professional skills and teachings of young people to ensure their stability in the labour market during the pandemic.

Research objectives:

1. to analyze the employment of young people, including during the COVID-19 pandemic;
2. to form a map of the transformation of the workforce and employment to determine the vector of professional development of modern youth;
3. to offer tools for the adaptation of young people to the labour market in a pandemic.

Research methods: theoretical analysis, systematization method, graphical method, economic and statistical analysis, systematic approach.

4 Results

Let's analyze the situation of youth participation in the labour force in the world and groups of countries by income level in dynamics (Table 1).

According to ILO (2020), even a smaller proportion of the working-age population receives income. This increases the likelihood that the redistribution of national income needed to ensure that everyone can enjoy a decent life will exceed the ability

Table 1 Labour force participation rate globally and by country group, 1994–2021 (%)

Country income group	Total		Youth (15–24 years)	
	1994	2021	1994	2021
World	65.4	60.3	56.4	40.7
Low income	74.0	70.5	62.6	55.4
Lower-middle income	60.3	55.1	47.8	34.5
Upper-middle income	71.0	63.4	65.1	42.4
High income	60.3	60.3	51.4	45.1

Source Compiled by the authors according to [14, 15]

of households to support their members [16]. At the same time, in the period 1994–2021, the reduction in the total population participating in the labour force was 7.8%. The decrease in employed youth over the same period in the world amounted to 27.8%.

The reasons limiting the employment of young people are the dilemma between employment and education, the low level of remuneration for young people, and, accordingly, low motivation for employment, employers' preferences in favour of specialists with work experience, as well as the choice of forms of employment by young people that are not recorded by official statistics (online employment, shadow employment schemes, freelancing, etc.) [30]. Thus, many young people prefer to receive professional and higher education, without combining this process with employment, thus investing in the development of human capital [8]. Indeed, the enrollment rate of young people in 1994 was 55%, in 2019 it reached 77%. This trend fully corresponds to the current stage of the formation of the digital economy and the development in the foreseeable future of a wide range of new professions and specialities [21]. As a rule, earning income and opening career prospects are the main motivating factors for youth employment. However, there is a dissonance between the expectations of young people and the low level of qualifications of young people, and, therefore, the low level of wages received (for example, in 2017, the earnings of employed youth in developing countries were less than $3 per day). A low level of income cannot stimulate young people to active employment, but such employment is an essential condition for moving up the career ladder. Hiring young specialists without initial work experience requires additional financial investments from the employer in the professional training of young specialists, their adaptation to the specifics of their activities, and also entails certain risks in the future [30]. Therefore, employers are more likely to give their choice in favour of specialists with work experience [5]. In addition, part-time employment, work in the informal sector of the economy that does not require high qualifications, or self-employment often becomes more attractive for young people [23].

As for youth employment, each country has its specifics of employment of young people by economic sectors. Thus, in Russia, according to the data of a sample survey of graduates of educational institutions of higher education in 2017–2019 by type of economic activity at the present or last job, the structure of youth employment is as follows [9]. In 2020, the share of employed graduates decreased by 10.5% compared to 2016. There have also been changes in the industry structure of graduate employment. In 2016, graduates preferred trade (13.6%), public administration (13.4%), and education (12.6%). In 2020, the share of employed graduates in healthcare and social services (12.8%), manufacturing (9.9%), information and communications (4.8%) increased against the background of a significant reduction in trade (12.4%).

A study of the sectoral characteristics of youth employment in the United States shows that 90% of Americans aged 16–24 are employed mainly in the private non-agricultural sector. 55% of all employed people aged 16–24 are registered in the leisure and hospitality industry (5.1 million people), in retail trade (3.8 million people), in the field of education and health services (2.1 million people). The manufacturing industry accounts for 6.5%. 1.3 million people are employed in the public administration sector. This distribution of employment reflects a long-term trend towards faster growth of employment in the service sector [28].

The pandemic has also had a significant impact on the state of the labour market and employment. In particular, in the months following the outbreak of the pandemic, there was a significant decrease in the employment rate of young people aged 16–24 years and a significant increase in the number of economically inactive young people. According to the ILO, unemployment among young people aged 15–24 in 2020 was 8.7% on average worldwide. This was followed by a smaller increase in unemployment [37].

For example, in the USA at the moment about 7.7 million young people are unemployed, about 3 million young people have lost their jobs over the past month [36]. Young people make up a significant proportion of people working in industries that are directly affected by COVID-19—retail and hotel business. Workers in these industries tend to be low-income workers, which increases their financial insecurity and vulnerability to job loss [35]. In the United Kingdom, 109% more young people applied for unemployment benefits by May 2020 than in March [37]. In general, in the European Union, the unemployment rate has increased significantly in the most affected countries—Italy, Spain, France. In Russia, the unemployment rate of youth aged 20–24 years increased in 2020 from 14.4 to 16.2%. However, to the greatest extent, the spread of the COVID-19 pandemic affected the unemployment rate in the most "prosperous" in the youth segment of the labour market, the age group of 25–29 years (an increase from 5.6 to 7.4%) [38].

The pandemic accelerated the transformation processes in the labour market, increased the requirements for the professional level of employees and updated new work models [36].Let's analyze the main trends in the transformation of the labour market and employment to determine the vector of professional development of young people and overcome the negative impact of the pandemic (Table 2).

A significant part of the net employment growth since 2005 has been in the categories of non-standard work (atypical employment) [33]. The growth of the platform

Table 2 Labour force and employment transformation map

Trend	The previous period	The current period	Transition difficulties
New models of work	Permanent job	Temporary work, part-time, independent contract work	Lack of management and legal protection of new labour models
Digital work design	Linear organizations with traditional services and classic employee interaction schemes	Exponential organizations with greater openness and the benefits of flexible security (flexibility and security). Democratization and de-hierarchization expand participation in decision-making at several levels: from team members and leaders to shareholders	Flexibility, speed and scalability requirements
Reskilling	The traditional approach to choosing an educational trajectory	Short-term retraining to meet the needs of the labour market is necessary to help more people acquire skills that meet the requirements of the labour market	The existence of a skills gap that requires active talent management strategies and sustained dialogue between companies, governments and educational service providers
Social protection	Traditional systems of distribution of responsibility for labour protection between governments and employers	Innovative regulation, consistent with the needs of employees and providing them with short-term reserves in case of periods of unemployment and outdated skills	Inequality in the labour market
Inclusive labour markets	Discrimination and lack of inclusiveness in the labour market based on age, gender, origin and other characteristics	Labour markets allow everyone to participate, regardless of race, ethnicity, age or origin, and facilitate the identification, assessment and validation of skills	Lack of initiatives to revise employment regulation and strengthen social protection systems

Source Compiled by the authors

economy has created more flexible opportunities for employees (in terms of prioritizing work-life balance, as well as ensuring immediate availability and increasing competition with permanent employees), as well as for employers (reducing the cost of permanent staff and rental payments) [36]. However, the existence of problems related to the lack of management and legal protection of contract work creates difficulties for the development of the institute of "independent workers".

The design of digital work is driven by the increasing needs of organizations for flexibility, speed and scalability. This requires new ways of organizing work both within firms and on their periphery. There is a transition from linear organizations to exponential ones, providing various working schemes for combinations of technologies, products and the market. For example, during the pandemic, the importance of global work in virtual teams was actualized. The new reality of work involves the transition from large structures to distributed smaller structures, from a hierarchical organization to structures based on teams and working groups, as well as a high level of trust. The design of digital work will be influenced by the greater openness of the company, the combination of the work of permanent employees with employees performing project work under a temporary contract, which will provide them with the benefits of flexible security and participation in decision-making at several levels.

Digitalization and demographic shifts require regular short-term retraining of employees to meet the needs of the labour market. Reskilling becomes the tool that will prevent the "displacement of a person by a machine". For example, Gartner predicts the creation of two million new jobs by 2025 based on the use of artificial intelligence. The lack of skills will highlight the growing gap between the complex tools that companies can produce with technology and the parts of these companies that can use the tools in production [1, 19]. For example, young people with ICT skills are perceived as "owning value" [2], since distance work increases the degree of social mobility of a person, helps to equalize his life chances, mitigates social inequality [31].

New work models and growing technological threats require the formation of social protection of employees based on innovative regulation, consistent with their needs. Such systems can provide employees with short-term reserves in case of periods of unemployment and outdated skills. The development of new social protection tools and new incentives can lead to an increase in guaranteed minimum income and ensure human dignity.

Technological disruption of labour markets creates not only some problems but also opens up new opportunities for people due to inclusivity—expanding access to labour markets for various categories of the population. The COVID-19 pandemic facilitated the transition of work to a virtual format, which had an impact on teamwork and interaction. New technologies (for example, blockchain) have created prerequisites for the development of new forms of entrepreneurship, and, accordingly, additional employment. This trend should be accompanied by a revision of employment regulation and the strengthening of social protection systems to increase the inclusiveness of labour markets [36].

Each country chooses its own set of tools to adapt young people to the labour market, as well as to support them in a pandemic. The proposed map of the transformation of the labour market and employment should become the basis for the implementation of a strategic vision for the formation of a set of measures of state policy in the field of employment.

Tools for adapting young people to the labour market in the context of a pandemic include:

- development of the institutional framework for the development of new forms of employment and expansion of employment channels (for state and business);
- stimulating entrepreneurial activity of young people (for state and educational organizations);
- flexible mechanisms of financial and other assistance to students to expand youth access to quality education (for the state, business, educational organizations);
- social, material, informational and psychological support for the most vulnerable youth (for the state, business, educational organizations);
- the creation of additional jobs for students and graduates; subsidizing employers when hiring or interning young professionals; tax benefits when hiring young professionals; temporary changes in labour relations that allow attracting young professionals to a wider range of jobs (for the state, business, educational organizations).

(1) National governments should consider the employment of young people, especially university graduates, as a priority state task and on an ongoing basis initiate laws and policy ideas that encourage young people to find employment, including within the framework of the implementation of new labour models. Significant measures to stimulate youth employment are: increasing the demand for labour; expanding employment channels [39].

(2) Young people aged 20–24 years demonstrate low economic activity, this is due to difficulties in finding employment and discrepancies between the requirements of employers and the expectations of future employees. One of the ways out of this situation may be the opportunity to open your own business. The task of educational organizations and the state is to increase the level of training of young specialists in the field of entrepreneurship, as well as the development of the infrastructure of entrepreneurship in the regions, in particular, business incubators to stimulate the entrepreneurial activity of young people [24].

(3) Some countries during the pandemic have proposed more flexible mechanisms of federal financial and other assistance to students to expand access to quality education: direct financial assistance to students and university graduates (France); an increase in the number of scholarships (Canada, Italy), the allocation of funds to help students who are in an emergency and need direct assistance (in the USA, $ 14 billion was allocated to universities for such support, in Germany, the German Student Union was allocated 100 million euros [4, 11]).

(4) Support for the most vulnerable youth is provided by countries taking into account new approaches from national governments. In particular, it can be direct financial support in case of dismissal in the form of a one-time social benefit, psychological support for young people. Information support for young people during the pandemic is provided thanks to accessible and open information from State institutions and social institutions [27].
(5) Some countries have taken measures to support young people in the field of employment to ensure the principle of inclusiveness of the labour market during the pandemic: the creation of additional jobs for students and graduates; subsidizing employers when hiring or interning young professionals (Canada, Germany); tax incentives when hiring young professionals (France); temporary changes in labour relations that allow attracting young professionals to a wider range of jobs (Spain, Canada).

5 Conclusion

Firstly, the authors analyzed youth employment on a global scale and by country groups. It was revealed that in the period 1994–2021, the reduction of employed youth in the world amounted to 27.8, (against the background of a decrease in the total population participating in the labour force of only 7.8%). The reasons for the identified reduction are disclosed. The industry specifics of the employment of young people in some countries have been identified. Secondly, a map of the transformation of the workforce and employment has been formed to determine the vector of professional development of modern youth. It reflects 5 key trends in the prospective development of the labour market—new labour models, digital work design, reskilling, social protection, inclusive labour markets. Thirdly, the authors have developed tools for the adaptation of young people to the labour market in a pandemic based on the proposed transformation map of the labour market and employment.

Data Availability

1. Data on Labour force participation rate globally and by country group, 1994–2021, are available in https://figshare.com/http:doi.org/https://doi.org/10.6084/m9.figshare.17000008.
2. Data on Labour force and employment transformation map are available in https://figshare.com/http:doi.org/https://doi.org/10.6084/m9.figshare.17000101.

References

1. Astera. (2017). By 2030, the market will need 298,600 specialists in machine learning and predictive analytics. https://astera.ru/news/?id=121226. Accessed: 15.10.2021.

2. Atayan, V. (2008). Axiological concepts of the regulatory function of value in society. *Humanities and Social sciences, 6*, 2–9.
3. Bonin, H., Gregory, T., & Zierahn, U. (2015). Übertragung der Studie von Frey/Osborneauf Deutschland [Transfer of the study from Frey/Osborn to Germany] (Mannheim, Center for European Economic Research (ZEW)).
4. Bundesministerium für Bildung und Forschung. (2020). [Federal Ministry oof Education and Research]. Coronavirus: was tut das BMBF? https://www.bmbf.de/de/coronavirus-was-tut-das-bmbf-11069.html. Accessed October 15, 2021.
5. Deloitte. (2015). The youth employment opportunity Understanding labour market policies across the G20 and Beyond. https://www2.deloitte.com/content/dam/Deloitte/de/Documents/public-sector/youth-employment-opportunity-labor-market-policies.pdf. Accessed October 15, 2021.
6. Dengler, K., & Matthes, B. (2015). Folgen der Digitalisierung für die Arbeitswelt: In kaum einem Beruf ist der Mensch vollständig ersetzbar [Consequences of digitalization for the world of work: There is hardly any job that people can be completely replaced], IAB-Kurzbericht No. 24/2015. Nuremberg, Institut für Arbeitsmarkt- und Berufsforschung [Institute for Employment Research].
7. Elhorst, J. P. (2003). The mystery of regional unemployment differentials: Theoretical and empirical explanations. *Journal of Economic Surveys, 17*(5), 709–748.
8. European Training Foundation (2015). Young people not in employment, education or training (NEET) an overview in ETF partner countries. https://www.etf.europa.eu/sites/default/files/m/BFEEBA10DD412271C1257EED0035457E_NEETs.pdf. Accessed October 15, 2021.
9. Federal State Statistics Service of the Russian Federation (2021). Human resources. https://rosstat.gov.ru/labour_force?print=1. Accessed October 15, 2021.
10. Gbadamosia, G., Evansb, C., Richardsonc, M., & Ridolfo, M. (2015). Employability and students' part-time work in the UK: Does self-efficacy and career aspiration matter? *British Educational Research Journal, 41*(6), 1086–1107.
11. Govinfo. (2020). The coronavirus aid, relief and economic security (CARES) Act. https://www.govinfo.gov/content/pkg/BILLS-116hr748enr/pdf/BILLS-116hr748enr.pdf. Accessed October 15, 2021.
12. Grushin, R. (2021). Problems of employment and youth employment in the context of the COVID-19 pandemic. *Young scientist, 10*(352), 82–85.
13. Gukasyan, Z., Tavbulatova, Z., Aksenova, Zh., Gasanova, N., Karpunina, E. (2022). Strategies for adapting companies to the turbulence caused by the covid19 pandemic. In Business 4.0 as a Subject of the Digital Economy. Springer, Switzerland.
14. ILO. (2020). World employment and social outlook: Trends 2020. International Labour Office, Geneva, ILO.
15. ILO (2020). Youth & COVID-19: Impacts on jobs, education, rights and mental well-being. International Labour Organization (ILO). https://www.ilo.org/global/topics/youth-employment/publications/WCMS_753026/lang--en/index.htm. Accessed October 15, 2021.
16. Karpunina, E., Galieva, G., Andryiashka, M., Vorobyeva, A., & Bakulin, O. (2021). Country risk assessment as a tool for improving the quality of state economic security management (on the example of Germany). *Quality—Access to Success, 22*(183), 136–142.
17. Karpunina, E., Butova, L., Sobolevskaya, T., Badokina, E., & Pliusnina, O. (2021). The impact of the Covid-19 pandemic on the development of Russian national economy sectors: analysis of dynamics and search for stabilization measures. *Proceeding of the 37th IBIMA Conference* (pp. 1213–1226). Cordoba, Spain.
18. Karpunina, E., Zabelina, O., Galieva, G., Melyakova, E., & Melnikova, Y. (2020). Epidemic threats and their impact on the economic security of the state. *Proceeding of the 35th IBIMA Conference* pp. (7671–7682). Seville, Spain.
19. Kergroach, S. (2017), Industry 4.0: new challenges and opportunities for the labour market. *Foresight and STI Governance, 11*(4), 6–8.
20. Khokhlova, M., & Khokhlov, I. (2017). Youth on the European labour market. *World Economy and International Relations, 61*(6), 48–56.

21. Klavdienko, V. (2019). Transformation of the employment structure of the population in the conditions of digitalization of the economy: Global trends and Russia. *Innovations, 10*(252), 81–87.
22. Klyachko, T., Loginov, D., Lomteva, E., & Semionova, E. (2020). Employment and features of youth employment during the pandemic. *Economic Development of Russia, 27*(12), 70–73.
23. Korchagina, I. (2019). Youth employment: Russian and foreign trends. *The Eurasian Scientific Journal, 5*(11). https://esj.today/PDF/51ECVN519.pdf. Accessed October 15, 2021.
24. Levina, E., & Dudin, M. (2020). The problem of youth employment in the context of global digitalization and application of network technologies. *Ekonomika truda [Labour economics], 7*(6), 519–536.
25. Lewis, P., & Heyes, J. (2017). The changing face of youth employment in Europe. Economic and Industrial Democracy.
26. Nataraj, S., Perez-Arce, F., Srinivasan, S., & Kumar, K. (2012). *What is the impact of labour market regulation on employment in LICs? How does this vary by gender?* RAND Corporation.
27. OECD (2020). OECD Policy Responses to Coronavirus. Youth and Covid-19: Response, Recovery and Resilience. https://read.oecd-ilibrary.org/view/?ref=134_134356-ud5kox3g26&title=Youth-and-COVID-19-ResponseRecovery-and-Resilience. Accessed October 15, 2021.
28. Petrovskaya, N. (2015). Youth in the American labour market. URL: https://rusus.jes.su/s207054760010126-5-1/. Accessed October 15, 2021.
29. RIA. (2020). One in five young people in the world lost their jobs due to coronavirus. https://ria.ru/20200527. Accessed October 15, 2021.
30. Sagina, O., Tavbulatova, Z., Perekatieva, T., Oganesyan, T., & Karpunina, E. (2020). Digitalization and employment problems of modern youth. *Proceeding of the 35th IBIMA Conference* (pp. 7692–7704). Seville, Spain.
31. Sheler, M. (1994). *Selected works*. Gnosis.
32. Sizova, I., & Khusyainov, T. (2017). Labour and employment in the digital economy: Problems of the Russian labour market. *Bulletin of Saint Petersburg State University. Sociology, 10*(4), 376–396.
33. Strebkov, D., & Shevchuk, A. (2009). *Freelancers in the Information economy: How Russians master new forms of labour and employment organization (according to the results of the first All-Russian census of freelancers)*. HSE.
34. Voronov, A., Sobolevskaya, T., Smirnova, E., Shugaeva, O., & Ponomarev, S. (2022). Managerial problems of enterprise development during the pandemic COVID-19. In towards an increased security: Green innovations, intellectual property protection and information security, Lecture Notes in Networks and Systems, Springer. pp. 841–851.
35. Weforum (2020). COVID-19: Young workers in the U.S. are likely to be hit the hardest. https://www.weforum.org/agenda/2020/04/young-workers-covid19-economics-united-states-service-industry-coronavirus. Accessed October 15, 2021.
36. Weforum. (2021). Workforce and Employment. https://intelligence.weforum.org/topics/a1Gb0000000LJQ4EAO?tab=publications&utm_source=sfmc&utm_medium=email&utm_campaign=2752036_Agenda_weekly-6August2021-0210804_095303&utm_term=&emailType=Agenda%20Weekly. Accessed October 15, 2021.
37. Youth Unemployment Statistics. (2020). House of Commons. Briefing Paper, No. 5871, June 16 2020.
38. Zabelina, O., Asaliev, A., & Druzhinina, E. (2021). Problems of the youth segment of the Russian labour market and novelties of the youth employment policy. *Ekonomika truda [Labour economics], 8*(9), 985–1002.
39. Zhang, Z. (2020). Particularities of youth employment in Russia and China. *Ekonomika truda [Labour economics], 7*(11), 993–1006.

40. Zubok, Yu., & Chuprov, V. (2012). Youth on the labour market: Transitive processes in the conditions of post-Soviet transformation. *Socio-Humanitarian Knowledge, 3*, 3–27.
41. Zudina, A. (2019). Do not work and do not study: NEET youth on the labour market in Russia. *The World of Russia: Sociology, Ethnology, 28*(1), 140–160.

Adaptation of Students to the Digital Space of the Modern World: Problems of Legal Support in Russia

Olga Y. Aparina, Irina N. Klyukovskaya, Irina N. Ter-Avanesova, Natalya R. Chernienko, and Irina V. Jacobi

Abstract The integration of the individual into a fundamentally new virtual socio-cultural environment, associated with the modernization of existing skills or the acquisition of additional digital competencies necessary for successful functioning and self-realization in a changing society, makes it possible to speak about the emergence of a cyber form of socialization of the individual "digital socialization". The authors identify the main directions of digital socialization and the main types of digital competencies acquired by users during the educational process using modern information technologies. The article analyzes the results of a specific sociological survey conducted by the authors, illustrating both the dynamics of the development of interests and needs of modern users of information educational resources, and problematic moments in the process of forming digital competencies. The compliance of the legal framework of the Russian Federationwith the prevailing needs and challenges of modern society in the context of the transformation of the modern institute of education is analyzed. The authors conclude that the growing demand for new sets of professional digital competencies in the modern labor market, which has changed under the influence of digitalization and automation of processes and the aftershock of the global pandemic, will inevitably entail further theoretical elaboration and correction of the understanding of the essence of professional digital competencies.

O. Y. Aparina · I. N. Klyukovskaya · I. N. Ter-Avanesova (✉) · N. R. Chernienko · I. V. Jacobi
North Caucasus Federal University, Stavropol, Russia
e-mail: iteravanesova@ncfu.ru

O. Y. Aparina
e-mail: apparina@mail.ru

I. N. Klyukovskaya
e-mail: ikliukovskaia@ncfu.ru

N. R. Chernienko
e-mail: magdalina2000@yandex.ru

I. V. Jacobi
e-mail: 4437@mail.ru

V. N. Ostrovskaya and A. V. Bogoviz (eds.), *Big Data in the GovTech System*,
Studies in Big Data 110, https://doi.org/10.1007/978-3-031-04903-3_19

Keywords Digital socialization · Distance education · Information technology · Personality adaptation · Digital competence · Pandemic education

JEL Codes I21 · K000 · K240 · K400 · O330 · O350

1 Introduction

> Growing up, learning and socialization take place in a hyperinformational society. The process of socialization through traditional institutions (families, schools) is increasingly supplemented by mass media and mass communications, especially the information and telecommunication network Internet, which are becoming the most important institutions of socialization, education and enlightenment of the new generation, to a certain extent replacing the traditionally established forms [13].

Globalization of the modern world and technological synergy, constant and continuous development and introduction of information technologies into all social relations without exception ensures exponential growth of the information array, together with increasing the requirements for the level, nature and content of the process of socialization of all participants in this process. The need to integrate a person into a fundamentally new information environment, coupled with the modernization of existing skills or the acquisition of additional digital competencies necessary for successful functioning and self-realization in a changing society, make it possible to speak about the transformation of the process of personal socialization and the emergence of its cyber form—"digital socialization". The study of this phenomenon has gained greater relevance in the context of the pandemic, which has come as a shock to established socio-cultural and communicative practices and has exacerbated the problem of the effective use of information and communication technologies in the most important areas of human life.

The analysis of scientific research in the Web of Science databases and devoted to various aspects of personality socialization (25,679 articles) shows a gradual but steady increase in the demand for the study of its digital form. If in 2002 we find only one mention of "digital socialization", then by 2019 their number increases to 126. Social distancing, objectively necessary during the pandemic period, naturally gave rise to a tendency to intensify the digitalization of all social mechanisms, which in turn established a "qualification" for the use of social benefits in the form of new skills and competencies. With an obvious lack of research in this area, it is necessary to highlight the works of Smith et al. [16] and Soldatova [17], which predict the possible nature of changes in the forms and mechanisms of traditional socialization when transferring it to a digital environment and the research of Attrill [1] and Stornaiuolo [19], which consider various aspects of the formation of a "digital personality". The works of such authors as Machackova et al. [12], Drozdikova-Zaripova et al. [4] are devoted to the problems of online modifications of certain types of deviant behavior and transformation of personality behavior models during unsuccessful socialization in a virtual environment. A number of such authors as Soldatova et al. [18] the concept of digital competencies is introduced as a criterion for the successful adaptation of various age groups of users in a virtual environment and their specifics are studied. Of considerable interest is the modern development of the theory of generations [9], according to which a further division of "Generation Z" or Digital Natives is carried out: by various types of professional adaptation [3], and the impact of digital communications on the formation of professional culture of young people is analyzed [6].

2 Materials and Method

The methodological basis of the study was the universal dialectical method of cognition and its conceptual provisions; general scientific and private scientific methods of cognition, comparative content analysis of digital media and law enforcement documents, analysis of statistical data were applied. At the empirical level of the study, a comprehensive concrete sociological study was conducted in the form of a survey of respondents. To determine the dynamics of the main indicators, the study was conducted in two stages: December 2020 and May 2021. The target group of a specific sociological study was made up of law students of a higher educational institution, who faced the need to adapt to the new conditions of the educational process, which is being transformed against the background of rapid digital development and restrictions caused by the COVID-19 pandemic. The choice of such a target group for conducting a case study is due to a study of the audience of active Internet users in Russia, which showed that 97.1% of them are young people under the age of 24 [10]. The survey results, along with a set of statistical data on the use of information technologies and the state of digital literacy in the world and the Russian Federation, as well as an analysis of the regulatory framework formed the empirical basis of this study. These issues were discussed at the round table "Information technologies in the higher education system—new challenges of the changing world", which took place within the framework of the week of the Department of Theory and History of State and Law of the Law Institute of the North Caucasus Federal University, Stavropol, May 27, 2021.

3 Results

The results of the study illustrate the dynamics of the development of interests and needs of modern users of information educational resources and the main directions of digital socialization of modern young people receiving higher education in today's conditions. The results obtained allow us to determine the characteristics of digital competencies acquired by users during the educational process using modern information technologies, and can also serve as a basis for predicting possible problems in the process of forming digital competencies. The study confirmed the basic working hypothesis: changing the mechanism of personality socialization in the socio-cultural digital environment allows us to identify the main directions of digital socialization of users: media-technical, communicative-recreational, functional-cognitive and professional.

Digital media-technical competencies, which are of primary importance, imply the formation of such basic components of this competence among users as a high level of digital literacy and technical awareness. When conducting a subjective assessment of the level of their competence in this area, the majority of respondents (75%) noted that they easily switched to a new format of distance learning without any difficulties in mastering the proposed technologies, at the end of the year 88% of respondents considered the problems insignificant. The number of students who have met difficulties on this path has not changed after a year, amounting to 10%. Despite the fact that the vast majority of respondents feel comfortable in virtual reality (the summary indicator was 67% in 2020 and 72% in 2021), only 8% have objectively specialized media and technical knowledge in the digital sphere as of December 2020 and 9% as of May 2021. An interesting fact is that the surveyed users tend to assess the level of digital literacy of their friends as quite high (27%) or equivalent to their own (56%), but the level of digital literacy of older people is considered low by the majority of young people surveyed (68%). Modern virtual space uses information networks as a tool of social stratification, which exacerbates the problem of digital inequality and makes it possible to determine the acquisition of these competencies as a priority direction of digital socialization. The solution of this problem is possible both by using the means of legal regulation: Articles 34, 36 of the Federal Law of December 29, 2012 N 273-FZ (as revised on 02.07.2021) "On Education in the Russian Federation" (with amendments and additions, entered into force on 13.07.2021) [5], provide for measures of material, technical and social support for students, including through the provision of digital teaching and educational tools; and through self-regulation methods: with the help of additional educational courses and techniques practiced at the level of individual social groups. Subjective assessment of one's digital awareness and mobility, as well as objective indicators of digital literacy form the basis of one of the aspects of adaptation to the new information space, being components of media-technical digital competence. Problems in socialization in this direction may be caused by technical difficulties, or lack of motivation to gain new knowledge, which complicates or makes it impossible for the user to realize and work in a digital environment. The main feature of the digital socio-cultural space is the movement of the most important social processes into the framework of the ubiquitous Internet—"a giant modeling process in which the most delicate communication mechanisms are used" [2] and an additional social sphere is formed—a virtual social environment, in which comfortable functioning is impossible without **digital communicative competence**. Successful socialization in the communicative and recreational direction increases the potential of a participant in digital interaction, expanding his ability to effectively realize himself in the virtual world and consists in the communicative strategies, both in direct and indirect forms, allowing to distinguish its components such as self-presentation, gaming and perceptual [21]. All modern users are active users of social networks, but the analysis of activity shows an interesting dynamic: if in 2020 respondents were inclined to pay their attention to various social networks (74% had one active account in several social networks, and 7% had several parallel accounts within one social network), then by 2021 the attention of the conditional "user" has focused on one selected network—this group of users has increased from 16 to 27%. This percentage has grown slightly, but against the background of the full coverage of the audience by social networks, this may indicate a purposeful utilitarian need for social networks in modern life. The priority activity in social networks for the majority of users (66% in 2020 and 51% in 2021) is communication with other users: chats, correspondence, exchange of comments and opinions. 23% in 2020 and 34% in 2021 used social media platforms to watch movies and listen to music. For educational purposes, to read news, information posts or articles, only 11% of students entered social networks in 2020 and 15% in 2021. The number of those who love games in social networks remained invariably zero. The communicative competence acquired in virtual reality retains its value in the physical world, which can additionally be recognized as one of the markers of the success of complex socialization, since the presence of difficulties with communication in the digital space can be an indicator of problems with socialization in the traditional communicative aspect. Problems or deviations along this path may consist both in the absence of a competent, balanced formation of self-presentation tactics, and in the risk of a violation of the perception of one's own identity. E. Toffler also anticipated the dangerous development of technology, predicting that in this case "the boundaries between the real and the unreal will blur, and society will face serious problems" [22]. With the latest digital XR (extended reality) technologies, the technique of immersion, the design of artificial reality or the use of augmented reality becomes possible, which on the one hand expands the cognitive and creative capabilities of the individual, but on the other - causes a number of reasonable concerns in the light of assessing the prospects of the impact of modern innovative technologies on the user's health or the nature of the perception of one's own identity, which focuses attention on the need to educate and develop critical thinking in a modern user aimed at objective cognition of the surrounding reality in order to preserve a sense of its significance. In order to guarantee the harmonious provision of this direction of digital socialization, it is worth noting the priority role of legal acts regulating such issues as ensuring confidentiality and data protection of all participants in the educational process, ensuring the digital security of users. Significant progress in the regulatory and legal regulation of this area has already been achieved. So, the provisions of Federal Law No. 273-FZ of December 29, 2012 (as amended on 02.07.2021) "On Education in the Russian Federation" (with amendments and additions, entered into force on 13.07.2021) [5] can be taken as an example. Paragraph 9 of article 13 of this law prohibits the use of educational technologies that harm the physical or mental health of students in the implementation of educational programs, methods and means of teaching and upbringing. But at the same time, there is no exhaustive and up-to-date list of information tools and educational techniques, and the long-term prospects of the influence of modern technologies on the personality are not only not studied, but also not predictable today. This is quite explained by the exponential development of modern technologies and again brings to the fore the mechanisms of self-regulation of the social groups involved, whose primary task in the process of digital socialization in the communicative and recreational direction is the formation of a culture of network communication as a basis for further safe, active and conscious perception of digital space.

In modern real virtuality, "technology" is not so much a physical set of machines and mechanisms, as a set of rational methods of mastering the world [7] and a virtual way to reveal the truth in the process of information interaction [8]. The formation of the ability to perceive and learn information in digital reality, based on the critical selection of the most effective and high-quality sources, is part of digital socialization and forms **digital functional and cognitive competence**. Difficulties in this area may consist in a complete or partial lack of interest in the process of cognition. Many researchers note that the use of the Internet by young people, which arose in the discourse of academic necessity, is more often utilitarian in nature [11], as a result of which users tend to limit themselves to information that gives a superficial idea of the issue being studied. The authors developing the "theory of generations" note the following main features of modern young people - Digital Natives ("Generation Z") as practicality, individualism and the ability to multitask against the background of difficulties in concentrating and increased suggestibility. Representatives of Generation Z have knowledge from various fields, but often superficial and unsystematic: they "believe that they know a lot, learn a lot, but cutting off Internet access almost instantly turns their knowledge only into opinions about knowledge" [15]. When conducting their own research, the vast majority of students turn to digital media, the Internet and electronic libraries (90% in 2020 and 87% in 2021). A slight decrease in the percentage of popularity of Internet sources may indirectly indicate cyber fatigue of students, but the priority still remains for digital media. The priority of digital technologies over traditional ones is an objective fact of modern reality. The issue is not the competition of paper and electronics, since the protection, preservation and support of "traditional and familiar to citizens (non-digital) forms and means of information dissemination is provided for in paragraph 3 of Article 1 of "The Strategy for the Development of the Information Society in the Russian Federation for 2017–2030" [20]. The problem lies in preventing unfair forms of competition in the new digital environment, for which Federal Law No. 149-FZ of July 27, 2006 (as amended on 02.07.2021) "On information, informational technologies and the protection of information" paragraph 8 of Article 3 establishes the inadmissibility of establishing any advantages of using some information technologies over others, unless the mandatory use of certain information technologies for the creation and operation of state information systems is not established by federal laws. Unfortunately, the widespread technological modernization and digitalization of learning and cognition processes does not entail the full use of all the potential provided. Despite the obvious popularity of digital content, the need to register in Internet libraries or on specialized portals deters 85% of users from this content, only 15% agree to spend time on registration. Internet sources in foreign languages have become the most unpopular and rarely used among modern students, especially if access to them involves the need to go through the registration procedure, only 2% of respondents in 2020 and 4% in 2021 noted that they consider them as material for their scientific research. The formation of the cognitive component of personality adaptation in a virtual socio-cultural environment, in turn, is a structural part of consistent self-realization and the basis for further socialization in a professional direction.

Obtaining **digital professional competencies** consists in the development of professional skills that mediate the commercialization of new technologies, both in existing employment areas and simulated ones. In the light of the ongoing social changes, only a full-fledged complex of both traditional and digital socialization of student youth, accompanied by the development of appropriate specialized competencies and the formation of interest and involvement in the development of professional skills in various fields of employment, reflecting the nature of the sphere of professional interests, is designed to ensure the relevance and subsequent creative and professional realization of the future specialist and should become the main criterion for the effectiveness of the functioning of the institution of digital education. Modern reality demonstrates the importance of professional digital competencies clearly. The unemployment rate in the Russian Federation last year showed an increase of 24.7% compared to the previous one and reached a peak of 4,321 million people, demonstrating an increased demand for workers with digital skills—vacancies involving "remote work" increased by 44% over the year, against the background of only 27% of Russians who have the necessary high level of mastering digital competencies. 49% of modern students noted the direct dependence of the success of further professional activity on the level of development of digital competencies, 36% found it difficult to answer this question. 15% of young people tend to underestimate the importance of digital professional skills. The formation of digital professional competencies, being a marker of successful digital socialization of an individual, as part of a general complex socialization is a complex multidimensional process. The request for transformation in this area inevitably entails changes at the level of regulatory regulation: the requirements of Federal State Educational Standards of Higher Education are dynamically transformed, which separately highlight the need for the formation of competencies in the field of information technology. In them, at the system level, a normative expansion of the list and content of professional digital competencies is carried out [14], amendments are made to the Federal Law "On Education in the Russian Federation" allowing educational organizations to form sets of competencies when drawing up educational programs. The demand for sets of professional digital competencies in the modern labor market, reformatted under the influence of digitalization and automation of processes, will inevitably increase the importance of methods for collecting and analyzing big data: analyzing an array of vacancies on career portals or interviewing specialists hiring for a vacancy that corresponds to a future educational program; the speed of reaction to the publication of vacancies or the general systematization of employment requirements in existing or completely new areas of employment will help clarify the blocks of digital competencies and focus on the availability of soft or hard-skills, if necessary, to form certain skill-sets.

4 Conclusion

Digital socialization as a complex process is of particular importance for people who make up the main audience of the virtual world - young people aged 16–25 years and consists in mastering the norms, rules and specific attitudes of the modern, virtualized socio-cultural space in several interrelated aspects: media-technical, interpersonal and professional, with the development of appropriate specialized competencies. The most important social institution regulating the process of socialization is an institution of education. It is the area that includes the formation of value orientations and normative attitudes of modern young people, effective comprehensive (including digital) socialization in a whole list of directions, as a result of which the student should develop a system of value orientations, form a stable antidestructive user immunity, form a culture of network communication, while maintaining a sense of the importance of the surrounding physical world and the integrity of his own personality. It is the Institute of Modern Education that is obliged to ensure that students successfully master the full range of digital competencies, which will be the basis for their further effective self-realization in the professional sphere. The syncretic nature of the problem defines the very essence of the implementation of digital socialization in the framework of future education, which consists in merging digital technologies and tools for the production of algorithms with a broad humanitarian education that develops students' emotional intelligence, using collective cultural practices that support joint creative activity and joint knowledge construction in the learning process, developing multifaceted digital competencies and accompanying soft-skills that guarantee full-fledged creative and professional realization.

References

1. Attrill, E. (2015). *The Manipulation of Online Self-Presentations: Create, Edit, Re-edit and Present*. Palgrave Macmillan.
2. Baudrillard, J. (2000). *Seduction* (p. 319). Ad Marginem.
3. Brodovskaya, E. V., Dombrovskaya, A. Y., Pyrma, R. V., Sinyakov, A. V., & Azarov, A. A. (2019). The impact of digital communications on Russian youth professional culture: Results of a comprehensive applied study. *Monitoring of Public Opinion: Economic and Social Changes, 1*, 228–251. https://doi.org/10.14515/monitoring.2019.1.11.
4. Drozdikova-Zaripova, A. R., Kalatskaya, N. N., Valeeva, R. A., Kostyunina, N.Y., & Biktagirova, G. F. (2019). Socio-psychological features of students inclined to victim behavior in the Internet. *Modern High-Tech Technologies, 12*(part 1), 159–166.
5. Federal Law of December 29, 2010 No. 436-FZ "On the Protection of Children from Information Harmful to Their Health and Development" (as revised on 01.07.2021). Consultant Plus. http://www.consultant.ru/document/cons_doc_LAW_108808/. Accessed 10 Oct 2021.
6. Gui, M., & Argentin, G. (2011). Digital skills of internet natives: Different forms of digital literacy in a random sample of Northern Italian high school students. *New Media and Society, 13*, 963–980. https://doi.org/10.1177/1461444810389751
7. Habermas, Y. (2007). Technology and Science as "Ideology". *Digest of Articles* (p. 208). Moscow, Praxis.

8. Heidegger, M. (1993). *Time and Being: Articles and Speeches* (p. 447). Republic.
9. Howe, N., & Strauss, W. (1991). *Generations: The History of America's Future, 1584–2069*. William Morrow and Company.
10. Internet audience in Russia in 2020. Mediascope research. (2021). https://mediascope.net/news/1250827/. Accessed 12 Oct 2021.
11. Kalmus, V., Runnel, P., & Siibak, A. (2009). Opportunities and benefits online. In S. Livingstone & L. Haddon (Eds.), *Kids online: Opportunities and risks for children* (pp. 71–82). The Policy Press.
12. Machackova, H., Pfetsch, J., & Steffgen, G. (2018). Editorial: Special issue on bystanders of online aggression. Cyberpsychology. *Journal of Psychosocial Research on Cyberspace, 12*(4). https://doi.org/10.5817/CP2018-4-xx.
13. Order of the Government of the Russian Federation dated 02.12.2015 N 2471-r "On approval of the Concept of information security of children". http://static.government.ru/media/files/mPbAMyJ29uSPhL3p20168GA6hv3CtBxD.pdf. Accessed 12 Nov 2021.
14. Order of the Ministry of Education and Science of Russia of 13.08.2020 N 1011 "On the approval of the federal state educational standard of higher education—bachelor's degree in the field of training 03.04.01 Jurisprudence" (Registered in the Ministry of Justice of Russia on 07.09.2020 N 59673) (ed. from 26.11.2020). http://publication.pravo.gov.ru/Document/View/0001202009070039. Accessed 10 Oct 2021.
15. Panov, V. I., & Patrakov, E. V. (2020). Digitalization of the information environment: risks, perceptions, interactions: monograph. Moscow: FSBI "Psychological Institute of RAO"; Kursk, "University Book", p. 199. https://doi.org/10.47581/2020/02.Panov.001.
16. Smith, J., Hewitt, B., & Skrbiš, Z. (2015). Digital socialization: Young people's changing value orientations towards internet use between adolescence and early adulthood. *Information, Communication and Society, 18*(9), 1022–1038. https://doi.org/10.1080/1369118X.2015.1007074
17. Soldatova, G. U. (2018). Digital socialization in the cultural-historical paradigm: A changing child in a changing world. *Social Psychology and Society, 9*(3), 71–80. https://doi.org/10.17759/sps.2018090308
18. Soldatova, G. U., Nestik, T. A., Rasskazova, E. I., & Zotova, E. Y. (2013). Digital competence of teenagers and their parents. Results of the all-Russian research. Internet Development Fund (p. 144). Moscow.
19. Stornaiuolo, A. (2017). Contexts of digital socialization: Studying adolescents' interactions on social network sites. *Human Development, 60*(5), 233–238. https://doi.org/10.1159/000480341
20. Strategy for digital transformation of the science and higher education industry. Ministry of Science and Higher Education. (2021). https://www.minobrnauki.gov.ru/documents/?ELEMENT_ID=36749. Accessed 12 Oct 2021.
21. Ter-Avanesova, I. N. (2021). Problems of "youth age" limits definition in the context of digital socialization. SEARCH: Politics. Social Studies. *Art Sociology. Culture, 2*(85), 49–57.
22. Toffler, E. (2002). *Shock of the future*. AST Publishing House.

Efficiency Assessment of Private Investors' Potential for Public–Private Partnership Projects

Svetlana B. Globa, Evgeny P. Vasiljev, and Viktoria V. Berezovaya

Abstract The purpose of the work is to study the issues of choosing a company for the implementation of concession projects and public–private partnerships, which has the greatest potential for these purposes. The authors believe that in order to optimize the use of limited state resources, an objective assessment of the effectiveness of joint projects at all stages of their life cycle is required, taking into account the achievement of a balance of interests of the parties (authorities, business structures, consumers), the impact of the results of these projects on the economy of the region as a whole. A reasonable assessment and management of the effectiveness of concession and public–private partnership projects will make it possible to determine the range of potential private investors interested in implementing such a project, taking into account the possibility of ensuring the required level of socio-economic efficiency.

Keywords Public–private partnerships · Infrastructure projects · Rating · Sustainable development · Private investor · Economic development · State

JEL Codes O44 · H54 · R58 · R11

1 Introduction

In conditions of limited financial and investment public resources, the problem of finding ways to manage the region's economy based on the interaction of government and business to solve the full range of tasks facing the state becomes urgent. According to world experience, the use of the partnership model in modern conditions is one of the most rational ways to organize cooperation between government and business, as it increases the interest of each party in the successful results of

S. B. Globa (✉) · E. P. Vasiljev · V. V. Berezovaya
Siberian Federal University, Krasnoyarsk, Russia
e-mail: sgloba@sfu-kras.ru

V. V. Berezovaya
e-mail: VVBerezovaya@sfu-kras.ru

V. N. Ostrovskaya and A. V. Bogoviz (eds.), *Big Data in the GovTech System*,
Studies in Big Data 110, https://doi.org/10.1007/978-3-031-04903-3_20

joint activities [3, 11, 12]. Public–private partnership, which allows attracting private financing, innovative technologies and business management experience to achieve long-term development goals of the territory, is actively used, first of all, in infrastructure industries [2, 5, 6, 8]. However, the potential of such a partnership is not fully used, due to the lack of a common understanding of its nature, the presence of organizational, institutional and economic constraints and barriers [7, 9, 10].

At the initial stage of the initiation of the concession and public–private partnership project, it is necessary to analyze the future project taking into account the specifics of the sphere in which it will be implemented; regional characteristics; the standard of living in the region and the impact on the social sphere; technical and technological characteristics and others. This will allow us to evaluate the effectiveness of the project for the correctness of the methodology and the reliability of the basic conditions laid down in it.

At the same time, it is proposed that the evaluation of the effectiveness of concession and public–private partnership projects should be carried out taking into account the following criteria, distributed by type of analysis.

1. Analysis of the financial model of implementation:
   - assessment of demand for products/services;
   - estimation of operating expenses;
   - income assessment;
   - assessment of private and public (municipal) financing;
   - assessment of the share of the equity and attracted capital in the volume of investments;
   - evaluation of commercial performance indicators for the project (NPV, IRR);
   - evaluation of the budget efficiency indicators of the project;
   - assessment of the sensitivity of project performance indicators with respect to changes in key variables;
   - assessment of the reasonableness of the project budget, etc.
2. Analysis of the social effects of the project implementation:
   - the value of socio-economic efficiency;
   - the impact of projects on social development indicators;
   - the level of tariffs for users (compared to the initial situation and similar facilities);
   - the level of quality of services provided;
   - the share of users' expenses for receiving services in the overall expenditure pattern, etc.
3. Project risk analysis:
   - the risk of increasing the cost of construction;
   - the risk of increasing the duration of construction;
   - operational risks;
   - financial risks;
   - market risks;

  - technical risks;
  - technological risks;
  - the risks that threaten the achievement of planned profitability;
  - the risks that threaten the achievement of planned technical and economic parameters;
  - social risks;
  - environmental risks.

4. Analysis of the project management system:
   - composition of the project team;
   - state regulation of the concessionaire's performance;
   - expenses on maintaining the management apparatus;
   - efficiency of the management system.

5. Analysis of the project control system:
   - available control tools;
   - effectiveness of control tools;
   - the possibility of independent control.

A reasonable assessment and management of the effectiveness of concession and public–private partnership projects will allow determining the range of potential private investors interested in implementing such a project, taking into account the possibility of ensuring the required level of socio-economic efficiency [1, 4].

2 Materials and Method

In order to identify, rank and select potential private investors interested in the implementation of the concession and public–private partnership project, taking into account the possibility of ensuring the required level of socio-economic efficiency, "The intelligent system for evaluating companies for concluding concession and public–private partnership contracts" has been developed based on creating a rating of companies using a mathematical apparatus, presented in the form of a database. The formed system is applicable for any territorial entities.

The intelligent system provides the following opportunities:

- identification of factors influencing the selection of companies for the conclusion of contracts of concession and public-private partnerships, allowing at the first stage to assess the possibility of their effective participation in regional projects;
- evaluation of the results of financial, economic and business activities of the applicant companies in order to create a rating of companies to identify the priority of participation in projects;
- formation of the final graphical and mathematical interpretation of the choice of effective companies.

The database includes:

- three tables with initial data (table "General selection criteria"; table "Quantitative indicators"; table "Performance indicators");
- four query tables (evaluation of qualitative data; calculation of coefficient indicators; calculation of points; calculation of the companies' rating).

The official websites "Federal State Statistics Service", "Department of the Federal State Statistics Service", the Legal system "ConsultantPlus", and other reliable sources applicable within the analyzed territory are used as an information base for filling tables with source data.

Tables with initial data for assessing the ranking and selection of a company for the purpose of concluding concession and public–private partnership contracts contain statistical information on socio-economic indicators divided into several blocks depending on the time interval.

The table with the initial general data selection criteria contains information on the selected criteria used in the calculation of the rating in the future.

The tables "Quantitative indicators" and "Performance indicators" are designed to calculate the rating indicator and represent a set of statistical data reflecting the effectiveness and attractiveness of companies. The table "Quantitative indicators" evaluates indicators from Forms 1 and 2 of accounting statements, for example: monetary funds, long-term and short-term loans and credits, profit from sales, net profit, etc. The table "Activity Indicators" evaluates the mechanism of raising funds and the current volume of activity.

The evaluation of the indicators presented in the blocks is made taking into account the calculation of the index and compliance with the criteria. At the beginning of the evaluation process, statistical data are accumulated reflecting the key socio-economic indicators of the companies participating in projects financed under the concession and public–private partnership. As a result, a rating assessment of companies is given, reflecting the most potentially ready companies in terms of projects financed under the terms of a concession and public–private partnership. Rating evaluation of companies is carried out on the basis of mathematical calculation of indicators and final rating points, as well as graphing the cluster potential of territories.

The proposed method of selecting an effective company for the implementation of the project acts as a mechanism in the study of applicant companies and has the following stages:

Stage 1—Selection of a group of companies according to the project criteria.

Stage 2—Selection of companies.

Stage 3—Entering the received data into the table "General selection criteria".

Stage 4—Entering data on Reporting Forms 1 and 2.

4.1—Calculation by companies.

4.2—Consolidated financial statements calculation (an additional reporting collection program has been created).

Stage 5—A brief financial analysis of consolidated financial statements.

Stage 6—Calculation of the rating of an effective company for the implementation of the project.

Let us describe the algorithm of the methodology used to calculate the rating.

Stage 6.1—Selection and description of indicators by blocks: qualitative data; quantitative data; coefficient data.

Step 6.2—Entering data by blocks into the MS Excel workbook on the "Model by indicators" sheet.

Step 6.3—Calculation of data on the "Model by indicators" sheet.

6.3.1—Calculation of Average Values.

6.3.2—Using the IF/OR/AND function for ranking companies by final scores and placing them according to the parameters "recommended"/ "recommended with a condition"/ "recommended with restrictions"/ "not recommended"/ "not considered".

6.3.3—Calculation of Indexes in MS Excel on the Sheet "Indexes-Rating Model".

Stage 6.4—Counting points for the company.

Stage 6.5—Calculation of the rating of companies to determine the effective company for the implementation of the project.

The rating indicators are grouped by blocks:

- quality data—maximum value of 45 points;
- quantitative indicators—the maximum value of 462 points;
- coefficient indicators—the maximum value of 206 points;
- industry indicators—maximum value of 16 points.

Based on the initial parameters, the database allows you to identify and evaluate factors that may affect the financial stability of the company, for example: the company's experience in implementing similar projects, debt to the budget and extra-budgetary funds, the fact that the company is declared insolvent (bankrupt) in accordance with the law, beneficiaries, etc. The calculation is performed semi-automatically, depending on the completed data set "Sheet No. 1". The final value of the indicator is formed in the table "Qualitative data".

The database makes it possible to evaluate indicators such as liquidity, profitability, turnover, solvency and others. These indicators provide an assessment of debt management, asset management, effective use of material, labor and monetary resources, as well as the liquidity of assets. The calculation is performed automatically; the final value of the indicator is formed in the table "Coefficient indicators".

Then points are automatically calculated for each company. The aggregate rating indicator of companies is determined automatically; the final value of the indicator is formed in the table "Rating of companies".

3 Results

The determination of the final indicator is based on the sum of all indicators for each company; a point gradation in the rating is formed. The estimation of the indicators presented in the blocks is made taking into account the calculation of the index and compliance with the criteria. At the same time, companies can be assigned the following rating.

Rating A (from 219 to 274 points)—Recommended companies—fully meet all legal requirements and function effectively. In accordance with this, they are allowed to enter into concession and public–private partnership contracts primarily without special restrictions in accordance with the requirements of legislation (federal and regional programs).

Rating B (from 164 to 219 points)—companies recommended with the condition—fully meet all legal requirements, but the functioning of the company is insufficient, since they have not scored enough points in the "Quantitative indicators" and "Coefficient indicators" blocks. They are allowed to enter into concession and public–private partnership contracts in the second place and a smaller amount of required allocations is possible, based on repayment, urgency and interest payment.

Rating C (from 110 to 164 points)—Companies recommended with restrictions—fully meet all legal requirements, functioning is relatively efficient, since the points in the blocks "Quantitative indicators", "Coefficient indicators", "Industry indicators" have been scored below the threshold values. They are allowed to enter concession and public–private partnerships contracts in the third place, and a smaller amount of required allocations is possible, based on repayment, urgency and interest payment with the provision of additional security and guarantees against default.

Rating D (from 55 to 110 points)—Not recommended companies—the performance of companies is at a low level of efficiency, since the points in all the blocks have been scored below the threshold values.

Rating E (from 0 to 55 points)—Not recommended companies—the companies' performance is inefficient, since minimum points in all the blocks have been scored; they are not considered for concession and public–private partnership projects.

When working with the database, the main form opens, allowing the user to go to any stage of forecast calculations by clicking the button. From any sheet, at the click of a button, you can return to the main form.

The calculation carried out for developers of the Siberian Federal District showed that for participation in concession and public–private partnership projects, the rating of companies is as follows:

- rating A corresponds to 3 developers: JSC TDSK, LLC USK Sibiryak, LLC Novy Gorod with final scores in the range from 225 to 274;
- rating B does not correspond to any developer;
- rating C corresponds to 3 developers: LLC Housing Initiative, LLC SDS-Finance, LLC SZ Kvartal with final scores in the range from 111 to 162;
- rating D corresponds to 14 developers and their share is 25% of all the considered developers of the Siberian Federal District;
- 33 developers correspond to the E rating and their share is 64% of all considered developers of the Siberian Federal District.

Thus, the developed methodology allows to choose effective companies for the implementation of the concession and public–private partnership projects. This technique can be applied to other territories as well.

In addition, to calculate the rating, it is worth evaluating not only financial statements, coefficient indicators and qualitative data, but also project models in order to give more accurate final estimates when conducting a rating analysis of the company for state support.

4 Conclusion

Thus, the purpose of the methodology and database is not only the collection, processing, systematization and storage of information, but also the use of information for making management decisions in the aspect of improving the instruments of regional innovation and investment policy, taking into account strategic interests and priorities, the development of concession mechanisms and public–private partnerships, stimulating the development of municipal infrastructure.

A more accurate identification and justification of the choice of a company for the purpose of concluding concession and public–private partnership contracts based on the rating of companies using mathematical apparatus will increase the transparency of the bidding procedure in concessions and public–private partnerships, unify the procedures for selecting a private partner, reduce the risks of violation of antimonopoly legislation, provide high-quality, timely solutions and high speed of project implementation.

Acknowledgements The research was carried out within the framework of the research grant of Krasnoyarsk Regional Fund of support scientific and technical activities on the topic "Development of models of financial support for investments in the utilities infrastructure of the region, taking into account the best Russian and world practices and features of the spatial and territorial development of the Krasnoyarsk Territory", No. CF-835, agreement on the procedure for targeted financing No. 226 dated 20.04.2021.

References

1. Ablaev, I. M., & Akhmetshina, E. R. (2016). The role of the public private partnership in the innovation cluster development. *Journal of Economics and Economic Education Research, 17*(4), 220–232.
2. Bacchini, F., Golinelli, R., Jona-Lasinio, C., & Zurlo, D. (2020). Modelling public and private investment in innovation. GROWINPRO Working paper, 6/2020.
3. Berezin, A., Sergi, B. S., & Gorodnova, N. (2018). Efficiency Assessment of Public-Private Partnership (PPP) Projects: The Case of Russia. Sustainability, 10, 3713. https://www.researchgate.net/publication/328307515_Efficiency_Assessment_of_Public-Private_Partnership_PPP_Projects_The_Case_of_Russia. Accessed 10 Oct 2021.
4. Boyer, E., Cooper, R., & Kavinoky, J. (2000). *Public-Private Partnerships and Infrastructure Resilience: How PPP's Can Influence More Durable Approaches to U.S. Infrastructure* (p. 32). Washington.

5. de Albornoz Portes, F. J. C. (2017). Alliances: An Innovative Management Model for Public and Private Investments. Case Study of Innovative Projects–Successful Real Cases (pp. 79–99). https://www.intechopen.com/chapters/56180. Accessed 10 Oct 2021.
6. Jomo, K. S., Chowdhury, A., Sharma, K., & Platz, D. (2021). Public-Private Partnerships and the 2030 Agenda for Sustainable Development: Fit for purpose? DESA Working Paper, 148. https://www.researchgate.net/publication/340309521_Modelling_public_and_private_investment_in_innovation. Accessed 10 Oct 2021.
7. Macdonald, S., & Cheong, C. (2014). The Role of Public-Private Partnerships and the Third Sector in Conserving Heritage Buildings, Sites, and Historic Urban Areas. The Getty Conservation Institute, Los Angeles.
8. Panteleeva, M., Surnov, D., & Senchukov, D. (2021). Conceptual model assessing complex the public-private partnership projects effectiveness. *E3S Web of Conferences, 263*(15), 05026. https://www.e3s-conferences.org/articles/e3sconf/pdf/2021/39/e3sconf_form2021_05026.pdf. Accessed 10 Oct 2021.
9. Public–private partnerships for skills development: A governance perspective (2020). European Training Foundation (p. 96).
10. Timchuk, O. G., & Kazantseva, I. I. (2021). Innovative technologies in the Russian construction industry as a factor of economic development. *IOP Conferences Series: Earth and Environmental Sciences, 751*, 012180. https://www.researchgate.net/publication/351375411_Innovative_technologies_in_the_Russian_construction_industry_as_a_factor_of_economic_development. Accessed 10 Nov 2021.
11. Timchuk, O. G., & Sokolova, L. G. (2020). Public And Private Partnership As A Tool For Innovation And Investment Stimulation. Selection and peer-review under responsibility of the Organizing Committee of the conference: Trends and Innovations in Economic Studies, Science on Baikal Session. https://www.researchgate.net/publication/347392896_Public_And_Private_Partnership_As_A_Tool_For_Innovation_And_Investment_Stimulation. Accessed 10 Oct 2021.
12. Yescombe, E. R., & Farquharson, E. (2018). Public-Private Partnerships for Infrastructure: Principles of Policy and Finance, 2nd Ed (p. 484). Elsevier Science.

Assessing the Value of Marine Environmental Projects Using the Scrubbing System

Igor V. Shevchenko, Svetlana N. Tretyakova, Natalia V. Khubutiya, and Natalya N. Avedisyan

Abstract The purpose of the research is to structure methodological approaches to managing project value assessment and to develop new recommendations for their application in companies of different fields of activity and management methods. Typical of our days, there is an urgent need for the implementation of global environmental projects, the importance of both personalization of the product and its environmental friendliness is growing, and the influence of the consumer is increasing. The missions of companies are changing, especially the shipping company in the field of environmental management of business processes. Achievements of strategic and tactical goals serve the implementation of new missions. It is possible to achieve strategic goals by implementing a policy of competitive advantage, creating value for consumers. Actual scientific approaches serve to support methods and tools for assessing generalized indicators of project activities, as a rule, involve averaging a set of indicators, which does not always meet the urgent needs of project management. The information technologies and mathematical apparatus used for this, as a rule, do not take into account the heterogeneity of parameters that cannot be generalized and expressed by a single indicator.

Keywords Value · Project manager · Shipping Company · Scrubbing system

JEL Codes B26 · D46 · G17 · L21 · L91

1 Introduction

In a rapidly changing dynamic business environment, there is a growing need for effective management of projects, programs and portfolios, building an effective

I. V. Shevchenko (✉) · S. N. Tretyakova · N. V. Khubutiya · N. N. Avedisyan
Kuban State University, Krasnodar, Russia
e-mail: dean@econ.kubsu.ru

N. V. Khubutiya
e-mail: exclusi@list.ru

V. N. Ostrovskaya and A. V. Bogoviz (eds.), *Big Data in the GovTech System*,
Studies in Big Data 110, https://doi.org/10.1007/978-3-031-04903-3_21

project management culture, value-oriented and strategic goals brings the business a competitive advantage [2, 4].

It is the benefits of a project or program of the Shipping company that is of real value. Few companies have implemented a benefits management process, and even fewer have done so successfully. Benefit management allows a Shipping company to increase its efficiency but requires a well-developed strategy, understanding of business processes and trends of the entire maritime industry. The category of value has come to play the most important role in modern project management methodology. This was in response to a demand for greater flexibility and value creation. The transition to value-oriented management was associated with a change in the paradigm of market relationships [10]. The traditional domestic approach, when Shipping companies have been making the same product for decades, optimizing production, has been replaced by an era of "cyclical demand".

The life cycle of ships has decreased, and the main role is now played by consumers—companies for the transport of goods and passengers, which manufacturers are trying to please in every way.

The advent of mass production in the era of the industrial revolution has shifted the focus from customer satisfaction to efficiency gains. As markets developed in the early twentieth century, shipping companies began to face increasing competition for market share (Greek, Turkish Shipping Companies), but they relied on advertising and promotion of goods and services, which aimed to impose products, goods (services) on small consumers. that they didn't want to buy [6, 9].

The buyer is no longer a recipient, not a mine for business; now the buyer is a co-manufacturer, co-engineer, co-designer of value creation. The project cannot fail to take into account the end-user. Customer value creation is a key priority for the project manager and cannot be transferred solely to the customer's responsibility once the product has been commissioned. Moving to value-based project management requires a deeper understanding of the term value and what it means. Behind the problem of values are complex issues of economic, legal, political and spiritual development of modern society. The problematic nature of the definition of the essence of the concept of value is determined, first of all, by its integral, specific features and characteristics. In modern literature, there are more than twenty definitions of this concept, where, in fact, various approaches and views on the problem under study are presented. The paradigm shift has shifted the focus of modern research in project management.

The most developed is the concept of value in the Japanese P2M project management standard. In P2M, the project is seen as a value-oriented measure that is based on a specific mission, carried out in a limited period and within the provided resources and external circumstances. In this case, value is considered as the fundamental principle of the project's existence: there is no value, no project. The Japanese standard defines the value of a project as well-balanced interests of all stakeholders.

2 Methodology

The first generation of project management methods was based on the quality-time–cost triangle, which was to answer the questions of efficiently allocating resources and completing planned tasks within a work hierarchy to achieve the unique goals of the project. Project management of the second generation expanded its view of the organization and management processes directly, the purpose of which was the formation of a project management system. Organizational structures with well-functioning internal communications, including the use of information systems, began to be developed. The third generation of project management theory proclaims itself the "Japanese school', reflected in the "P2M" standard [1, 8]. She views the project as integral to the organization that implements it, but also within the entire environment of the project. Emphasis is placed on the mission and value of the project, which is brought into the environment in which the project exists. It is important to use the results of the project, the exploitation of the value that was obtained while working with knowledge, skills and abilities. Any project should carry out some mission related to the creation of something new, unique. In everyday life, business and social interaction, value creation turn into a constant search for the optimum point. Value creation relates to the needs of a market, industry or industry, which can be addressed by one or more people using intellectual, physical or financial resources.

The formation of market relations in the economy of the state led to the emergence of completely new independent directions in management, which were developed as a result of a critical rethinking of modern foreign theory and practice of management science, the creation of unique management approaches, methods and tools. The most significant place in the structure of modern management is occupied by the theory of project, program and project portfolio management.

Recently, there has been a constant increase in the needs of various organizations and companies for the implementation and use of project management principles and systems. Project management has gained recognition all over the world and has become one of the main management tools both in the market economy and in the field of public administration. In this regard, the number of research and consulting organizations that develop, implement and use modern approaches to project management has increased. All this testifies to the fact that the branch of project management, which until recently was considered as a completely new branch of knowledge and skills has become firmly established in modern management. Value creation and management is a collection of management tools and applications (models, methods, approaches) that allow you to create and manage value [7, 9, 10].

3 Results

The PMI's A Guide to the Project Management Body of Knowledge (PMBOK), a recognized international standard in this area, describes the project as a stopgap measure designed to create unique products services or results.

Project management in modern management is one of the most effective tools actively used in various areas of business, in particular in heavy industry. The project, which is based on three basic postulates temporal nature, unique character and a significant indicator of uncertainty, implies the use of special tools that make it possible to effectively manage its components and achieve the set goals. Project management methodology is the formalized principles, rules and processes of project management. The methodology consists of several levels: international project management standards, local project management standards, regulations on project activities in the organization, and others.

The main international standards of project management today are ISO 10006: 2017, Quality management systems—Guidelines for quality management in projects, PMBOK—the standard of the American Institute of Project Management PMI, PRINCE2 (Projects In A Controlled Environment)—the standard is approved by the UK government, P2M—Japanese project management standard, GPM ® Global P5 TM—P5 standard GPM ® Global P5 TM standard—UN initiative, Agile Manifesto—flexible project management manifesto, etc.

Value expressions of indicators of business activity as the total value of the enterprise, the totality of all material and non-material subjects of the process. So, for example, material components—cash, capital, fixed assets and information. Key examples of intangible assets: assessment of the company's reputation, brand popularity, the level of public welfare, and others. Depending on the organization, the meaning of the expression "business value" can be reflected in the short, medium and long term. The value can be created through effective management of regular operations. However, through the effective application of RFQ management tools, organizations get the opportunity to apply robust processes to achieve strategic goals and business expansion the value of their investment. Of course, not all organizations are focused on generating super-profits associated with the business. The whole set of enterprises—state, private, commercial or non-profit—focused on bringing their activities to business value [7, 9, 10].

The cost approach of an environmental project considered in the analysis is based on the hypothesis that having established a system of indicators reflecting the growth in the value of an enterprise, we will be able to make a decision on participation. The variables most commonly used in calculations—net profit and profit growth—are not always suitable for this. Setting non-financial metrics such as customer satisfaction, product innovation, employee satisfaction, and other components becomes an important precondition. However, it is not possible to directly affect the cost of a project, so a valuation-based approach is a comprehensive understanding of the variables that are involved in creating value, namely the main drivers of value (Fig. 1). Value Drivers are built into indicators, which are usually divided into different levels of governance.

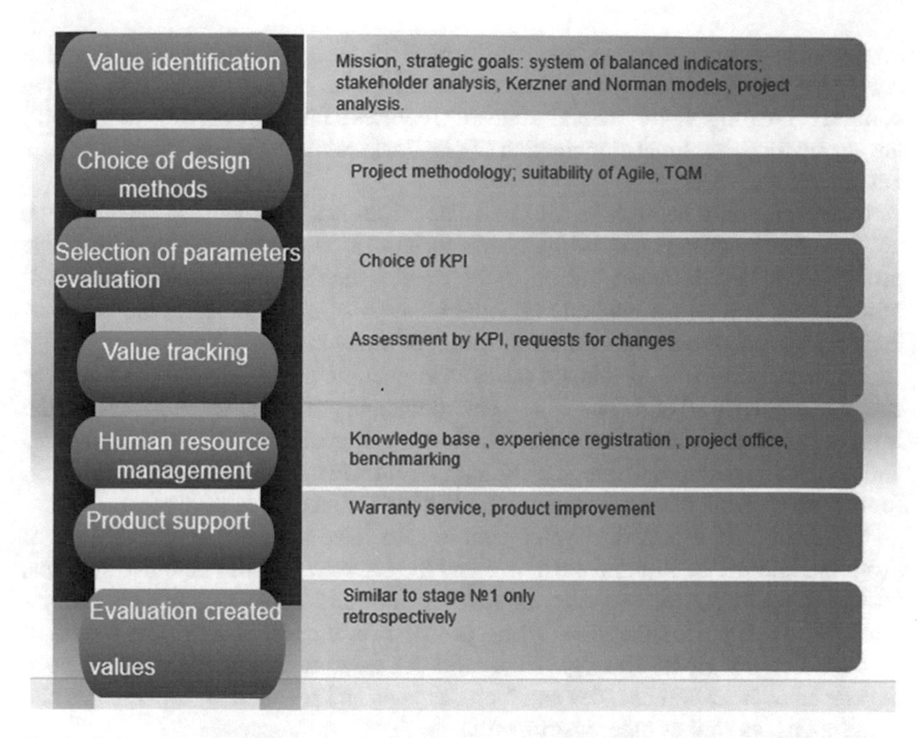

Fig. 1 Steps and tools for value management. *Source* Compiled by the authors based on [9, 10]

In terms of its content, it resembles a hierarchy, that is, a decomposition of key goals is displayed. The principle of oneness and complementarity provides line managers with controllable variables that they can regularly measure and work to improve. Of course, increased productivity affects the achievement of key higher-order goals and allows increasing the value of the project and the company.

Despite significant differences in the types of projects and a variety of conditions for their implementation, the assessment of the cost of projects and their expertise should be carried out based on the principle of uniformity—based on uniform reasonable indicators (variables) [7, 10]. Such variables are reflected in the following groups: methodological, or generalizing, ensuring the rational behaviour of customers, performers, regardless of the nature and goals of the project, methodological, providing an economic justification for assessing the cost of projects and decisions made on them. Key operating indicators, which should simplify the process of project cost estimation and ensure the required accuracy, are considered by us in other studies.

The cost approach as a set of tools for decision-making in company management, first applied in the late twentieth century, was a response to the need to establish a link between business goals and environmental goals. In modern realities, it is not possible to consider a company excluding its environmental awareness. Value management as a tool for integrating a company into the global ecological order is extremely important in the modern economic structure [5]. It is the multifactorial nature of the problems, a fundamentally new view of the enterprise management process in the context of creating environmental value that requires a modern company to intensify the search for the optimal combination of classical tools and the introduction of new management models.

Companies must be managed and make decisions in such a way as to create value not only for the owners but for the whole society [3, 5]. Increasing attention to the problem of the environment, the search for social justice has entered the global social order, going beyond national and regional boundaries. All businesses are aware of their role in the global environmental culture of doing business.

The habitual directions that we are used to using in the concept of development can increase the level of resources and environmental pollution; second, increase the level of transparency and social behaviour. Firms can create and test new technologies and production methods that reduce the harmful human impact on planetary ecology; fourth, the creation of a system of equitable distribution of wealth (Fig. 2).

The ecological aspect of development is associated with the impact of a set of programs and projects on the biosphere and the effective use of inanimate systems, including water and ecosystems, as well as flora and fauna that inhabit ecosystems. The biosphere is the place where people live; it can consist of the streets of cities, towns or regions. The legal framework for environmental protection has been developed over several decades and includes laws and regulations, agreements and conventions, as well as other instruments.

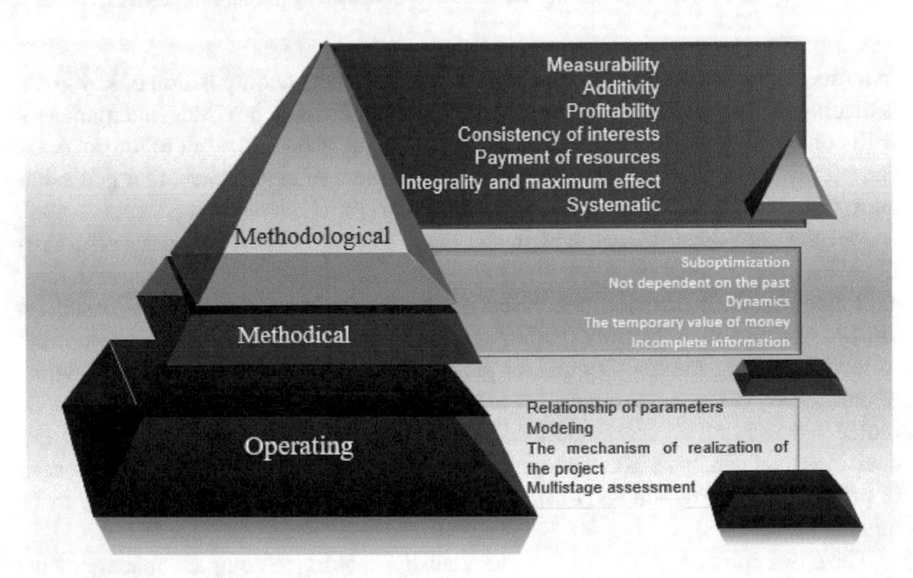

Fig. 2 Pyramid of assessing the cost of environmental projects of domestic companies. *Source* Compiled by the authors based on [7, 8]

The main question that the investor wants to get an answer to at the stage of preliminary examination of the project is undoubtedly the following: do the positive results of the project outweigh the costs of its implementation? In a shipping industry where small, incremental savings could mean the difference between being competitive or losing money, the scrubber alternative could make a big difference. All EGCSA member companies have provided scrubbers to the shipping industry and have had significant successes. As an example, CR Ocean Engineering (CROE) has several onboard units in successful operation, and many others in the design and fabrication stage for some very reputable shipping/cruising companies.

According to the second scheme, exhaust gases must be constantly monitored when the equipment is in use and there is no need for system performance certification. According to both schemes, the state of any wash water discharged into the sea must be constantly monitored for acidity, turbidity and PAHs (a measure of harmful oil constituents), and data must be recorded depending on the time and location of the vessel. A nitrate test is also required at every renewal certification.

Scrubbing Systems' utilization rates could be improved if flag states and other states offer government aid or attractive financial deals. So far, aid has been limited to a small number of projects in Europe and no projects in the CIS at all, and some financial institutions have begun to offer schemes that help with capital expenditures and that link disbursements to savings received [3, 5].

Simple arithmetic means averaging the boundaries. In particular, the width of the interval corresponding to the consensus (group) estimate obtained taking into account the width of the averaged intervals is equal to or less than the width of the group estimate obtained from the arithmetic mean.

In addition, the centre of the interval of the consistent group estimate based on the width-weighted averaging is always biased towards the centre of the less wide of the averaged intervals. Thus, if the width and centre of the averaged interval are interpreted as the quality of the expert assessment (the accuracy of the group assessment, the final assessment of the qualifications of experts), then averaging based on the ratio gives better (more accurate, more qualified) results of expert assessments.

A project charter is one of the first officially recognized project documents that enable a leader to start a project in an organization. By definition, a project should draw on important and relevant characteristics of the project. Teaching the working team the "basics" of the project is a kind of conviction (guaranteed) of the organization in the need to implement the project. The objectives of the initiation phase are recognition of the need for project implementation; identification of risks and objectives of the project; defining the principles of implementation; predicting outcomes from management and other stakeholders and identifying a responsible person. At the stage of transformation into life or "primary start"—the Charter (Passport) of the project, risks are calculated and a decision is made to launch a specific project by the manager. It is more practical to appoint people responsible for project implementation and risk calculations in advance. In other words, a responsible person should always be appointed at the beginning of planning and preferably at the stage of developing the project charter. A criterion for the effectiveness and failure of a project is a set of variables that make it possible to assess the success of the project and its cost. A project management plan is a project-forming consolidated document, in which we can find institutional subsystems, development scenarios and methods for calculating risks when making a decision. When developing a management plan, the project uses the results of other planning projects (including quantitative and strategic) to create a single document that will be the communication plan as a guide for the implementation and project of the project.

4 Conclusion

As part of the implementation of environmental projects, one should be guided by the "Express-analysis" method of the project. Basically, a similar technique is used at the initial stage of the "inception" of the project and is carried out by the initiator of the project. The set goal of the project, as in our study, the introduction of an ecological fuel purification system, allows us to solve several fundamental problems. Firstly, we offered a completely new way of deciding on choosing an ecological project. Secondly, the introduction of the methodology for conducting an environmental project in companies will allow the majority of domestic enterprises to quickly become involved in the global culture of environmental business conduct. It should be noted that the "Express-analysis" of the project includes a preliminary assessment, and cannot serve as a categorical criterion in choosing a project, but at the same time, it is the proposed methodology that enables those in charge to reveal the strengths of the project proposals.

References

1. Akhmedova, M., Egorov, M., Egorova, L., & Barishnikova, V. (2020). Methodological evaluation tools of world, national and corporate intangible assets as the basis for competitiveness management of companies. Lecture Notes in Networks and Systems: International Conference on Comprehensible Science, ICCS 2020, Carthage, Tunisia, 30 October 2020–31 October 2020: Book Series, 186, (pp. 551–559).
2. Khalafyan, A., Shevchenko, I., & Tretyakova, S. (2019). Data analysis as a toolkit for construction and evaluation of consistency of banks ratings. In Advances. Intelligent Systems and Computing: Book Series: International Conference on Digital Science (pp. 129–136).
3. Litvinsky, K., & Aretova, E. (2020). Mathematical modeling of optimal financing of investment modernization projects for fuel and energy complex enterprise. Lecture Notes in Networks and Systems: International Conference on Comprehensible Science, ICCS 2020, Carthage, Tunisia, 30 October 2020 31 October 2020: Book Series, 186, (pp. 31–53).
4. Malakhova, T. S., Dubinina, M. A., Maksaev, A. A., & Fomin, R. V. (2019). Foreign trade and marketing processes in the context of sustainable development. *International Journal of Economics and Business Administration,* 7(S1), 195–202. Piraeus, Greece, International Strategic Management Association (ISMA).

5. Perova, A. E., Satsuk, T. P., Slavin, A. M., Progunova, L. V., Veynbender, T. L., & Sokolova, G. N. (2018). International standards of the public sector financial reporting in ensuring economic security. *Revista Publicando: Scientific journal, 5*(18–2), 330–340.
6. Ratner, S., Chepurko, Y., Drobyshecskaya, L., & Petrovskaya, A. (2018). Management of energy enterprises: energy-efficiency approach in solar collectors industry: The case of Russia. *International Journal of Energy Economics and Policy (IJEEP): The International Academic Journal, 8*(4), 237–243. Mersin, Turkey, Econ Journals.
7. Ravindran, N., & Hasini, H. (2018). Experimental analysis of SMART scrubbing system with the principle of dynamic precipitation. *International Journal of Engineering, 7*(4), 317–322.
8. Shevchenko, I., Puchkina, E., & Tolstov, N. (2019). Shareholders and managers' interests: Collisions in Russian corporations. *SE Economic Journal, 23*(1), 118–142. Moscow, Russia, Publishing House of the Higher School of Economics.
9. Shevchenko, I., Tretyakova, S., Avedisyan, N., & Khubutiya, N. (2020). IPO–The Pattern of Hierarchy with a Variety of Alternatives upon Criteria. Economics, Business and Management Research: Proceedings of the 5th International Conference on Economics, Management, Law and Education (EMLE 2019), October 11–12 (pp. 91–96). Paris, France, Atlantis Press.
10. Tolstov, N. (2020). Model for assessment of the quality of financial management in solving the problem of agency conflicts. Lecture Notes in Networks and Systems: International Conference on Comprehensible Science, ICCS 2020, Carthage, Tunisia, 30 October 2020–31 October 2020: Book Series, 186 (pp. 91–100). Cham, Switzerland, Springer Science and Business Media Deutschland GmbH.

Diagnostics of Budgetary Potential of Regions in Order to Implement the Value-Oriented Financial Policy of State

Nadezhda I. Yashina, Oksana I. Kashina, Nataliya N. Pronchatova-Rubtsova, Sergei N. Yashin, and Victor P. Kuznetsov

Abstract The study is devoted to the development of theoretical and methodological foundations of value-oriented state budget management. The goal is to develop a methodological toolkit for diagnosing budgetary potential of regions in order to implement the value-oriented financial policy of state. The proposed toolkit is based on the use of scientifically based methods for determining budgetary potential, taking into account the economic development of regions to achieve high standards of quality of life of citizens. Economic and statistical methods were used to diagnose the budgetary potential. Approbation was carried out on the data of the Federal State Statistics Service and the Ministry of Finance of the Russian Federation for 2019. Results. The budgetary potential of the regions was assessed, the classification of the levels of budgetary potential was carried out on the basis of the sustainability index. It was found that the budgetary potential of the region is sensitive to its economic development and budgetary sustainability. Conclusions. The implementation of the methodology allows for a targeted value-oriented state financial policy in the regions. Monitoring and analysis of results determine effective government decisions to improve the quality and accessibility of education, in particular, training highly qualified personnel, stimulating the development of human capital as the main value of state. Regions with a consistently high economic potential based on industrial development have high budgetary potential.

Keywords Value-based approach · Budget management · Regional budget potential · Financial policy · Sustainability index

JEL Codes C51 · H61 · O21

N. I. Yashina (✉) · O. I. Kashina · N. N. Pronchatova-Rubtsova · S. N. Yashin
Lobachevsky State University of Nizhny Novgorod, Nizhny Novgorod, Russia
e-mail: sitnicof@mail.ru

V. P. Kuznetsov
Minin Nizhny Novgorod State Pedagogical University, Nizhny Novgorod, Russia

V. N. Ostrovskaya and A. V. Bogoviz (eds.), *Big Data in the GovTech System*, Studies in Big Data 110, https://doi.org/10.1007/978-3-031-04903-3_22

1 Introduction

Economic relations are subject to the complex impact of many factors, this determines the emergence of new threats and risks, and, consequently, the high variability of the economic environment, the natural complication of tasks and the need to search for non-standard ways to solve them in the field of state socio-economic policy.

Value-oriented budget management is the main tool for ensuring the social well-being of the population in the face of economic transformation and external challenges such as stock crises, financial sanctions and the pandemic [5]. The value-oriented approach is understood as a method of public finance management in which the achievement of a goal is assessed from the standpoint of universal human values, such as truth, goodness, socially and personally significant values in life. The value-based approach to public finance management allows timely provision of public funding in the field of health care, education, culture and other socially significant areas, and making management decisions that will contribute to the prosperity and well-being of citizens [1, 4, 7].

The development of methodological tools for diagnosing the budget potential of regions as a way to ensure value-oriented budget management is an important problem in the context of financial instability and crises.

The gap between Russian regions in the development of resource economic potential slows down the development of the entire country. Differentiation cannot but affect the state of public finances, this contributes to the growth of socio-economic inequality and the decrease in the investment attractiveness of regions [3].

2 Methodology

In the course of the study, the authors have formed the methodological toolkit for assessing the budgetary potential of the regions, which is the basis for the implementation of the value-oriented financial policy of the state.

1. At the first stage, quantitative indicators were formed, the analysis of which should lead to some qualitative conclusions about the budgetary potential based on the resource industrial and economic potential [9] and fiscal sustainability of the regions [10].
2. The second stage consists in the analytical processing of the collected data and the creation of a special system of indicators, including a set of indicators responsible for the level of resource industrial and economic potential (P1–P11), and a set of indicators characterizing the budgetary sustainability of the regions (P12–P25).

The characteristics of the indicators are presented in Table 1.

The designations in the table are deciphered as follows:

ND—tax revenues (income), NND—non-tax receipts (income), D—total budget revenues (revenues), FP—financial aid, GD—government debt, PR—the amount

Table 1 Systems of indicators for determining the budgetary potential of the constituent entities of the Russian Federation

No	Indicator name	Calculation formula	Interpretation of the indicator and requirements for its value
1	2	3	4
P1	Financial independence ratio	(ND + NND)/D	Shows how much the budget depends on financial assistance from the higher budget, or what share falls on the budget's own revenues; maximization of indicator
P2	Financial stability ratio	ND/D	Shows the share of tax revenues in total budget revenues, i.e. how many taxes go to the budget; maximization of indicator
P3	Ratio of own income and financial assistance	(ND + NND)/BP	Shows the ratio of own funds to financial assistance; maximization of indicator
P4	Coefficient of provision of financing of social expenditures with own income	(ND + NND)/SR	Shows how much the expenses for the social sphere are covered by the own budget revenues, i.e. to what extent the regional authorities can independently finance the social sphere; maximization of indicator
P5	Coefficient of provision of financing of industrial sectors with own funds	(ND + NND)/HE + ZKH	Shows to what extent own incomes can provide financing for production sectors, i.e. to what extent the regional authorities can independently finance the production sector; maximization of indicator
P6	Funding ratio for administrative expenses	OV/R	Shows the share of expenses for administrative needs in the total budget expenditures, i.e. how much money is spent on financing the management apparatus; minimization of indicator

(continued)

Table 1 (continued)

No	Indicator name	Calculation formula	Interpretation of the indicator and requirements for its value
P7	Human capital investment ratio	SR/R	Shows the share of social spending in total spending, i.e. how much money is spent from the budget to ensure the development of the social sphere; maximization of indicator
P8	Interest expense ratio	PR/SD	Shows what share of the cost of servicing municipal and state debt is covered by own revenues; minimization of indicator
P9	Total debt ratio	GD + PR/SD	Shows the share of borrowed funds in the total volume of own budget revenues, i.e. stability of the territorial budget; minimization of indicator
P10	Coefficient of diversion of public resources	PR/SR	Shows the proportion of budget expenditures on debt service to the amount of socially significant expenditures of the regional budget; minimization of indicator
P11	Borrowing structure ratio	KR/Z	Maximization of the indicator, so the cost of servicing loans is higher
P12	Median wage coefficient	ZPm/ZPsr	Maximization of the indicator, the ratio of the median wages and the average one
P13	Population	N	Maximization of the indicator, the size of the entire population
P14	Working age population	N TR	Maximization of indicator
P15	GDP per capita	VVP/CH	Maximization of indicator
P16	Average salary	ZPsr	Maximization of indicator
P17	Median wages of employees of organizations	ZPm	Maximization of indicator
P18	Fixed assets cost	OF	Maximization of indicator
P19	Fixed capital investments	I	Maximization of indicator

(continued)

Table 1 (continued)

No	Indicator name	Calculation formula	Interpretation of the indicator and requirements for its value
P20	Balanced financial result	FR	Maximization of indicator
P21	Turnover of enterprises	O	Maximization of indicator
P22	Cost of fixed assets/Turnover of enterprises	OF/O	Maximization of indicator
P23	Fixed capital investments/Turnover of enterprises	I/O	Maximization of indicator
P24	Balanced financial result/Turnover of enterprises	FR/O	Maximization of indicator
P25	Investment in fixed assets/Population	I/N	Maximization of indicator

Source Developed and compiled by the authors

of interest payments allocated to service the debt, R—budget expenditures, ZDN—arrears on taxes and fees, SR—social budget items, NE—spending on the national economy, ZKH—maintenance costs of housing and communal services, OF—cost of fixed assets, O—turnover of enterprises, I—fixed capital investment, FR—balanced financial result.

Budgetary sustainability of regions depends on volume and structure of sources of budget financing, as well as the amount of excess of revenues over expenditures on state reserves.

3. At the third stage, standardization of indicators, depending on their economic meaning and impact on budgetary potential, is carried out to conduct a comprehensive assessment [8].

For standardization while minimizing indicators (1) and for standardization while maximizing indicators (2), the following formulas are used.

$$K_{ij}^{*} = \frac{K_{ij} - K_{imin}}{K_{imax} - K_{imin}} \quad (1)$$

$$K_{ij}^{*} = \frac{K_{imax} - K_{ij}}{K_{imax} - K_{imin}}, \quad (2)$$

where K_{ij}^{*}—standardized indicator of the *i*-th proposed indicator for diagnosing the budget potential of regions in the *j*-th region,

K_{ij}—the calculated value of the *i*-th proposed indicator of diagnostics of budgetary potential of the regions in the *j*-th region,

K_{imax}—maximum of the *i*-th indicator, K_{imin}—minimum of the *i*-th indicator. Standardization builds all indicators from 0 to 1.

4. The ranking of regions is based on the formation of a comprehensive standardized indicator ($KKBP_j^{st}$), which is determined by the formula (3):

$$KKBP_j^{st} = \sum_{i=1}^{n} K_{ij}^{*}, \tag{3}$$

where $K_{ij}{}^{*}$—standardized value of the *i*-th indicator from the system of relative coefficients of budgetary potential in the *j*-th region.

The level of stability of budgetary potential is determined based on the optimality index. Optimality index calculation:

1. The regions are ranked and the rank is determined by the value of complex standardized indicator.
2. The region with the lowest value of the complex standardized indicator is assigned the first place or one point.
3. All ranks of all regions are summed up.
4. The index of sustainability of the budgetary potential is defined as the ratio of the rank of the corresponding region to the sum of all ranks of the region multiplied by 100%.

The index of stability of budgetary potential from 0.5 to 1.0 indicates a high level of stability and development of budgetary potential, shows the current trends in the distribution of tax revenues and the formation of tax base for the main types of taxes.

The main types of income are income tax, personal income tax, property taxes, etc. A slight increase in borrowing is possible.

The value of the index of stability of budgetary potential up to 0.5 indicates a satisfactory level of budget stability, characterizes the persistence of the current trends in the distribution of tax revenues and the formation of the tax base for the main types of taxes.

The main types of income are MET, VAT, personal income tax, taxes on small businesses. High borrowing growth is possible. Urbanization and agglomerative construction in Moscow and St. Petersburg cause difficulties due to serious structural changes in the capital markets, real estate, labor resources, which is reflected not only in their territorial development, but also in the development of regions.

The value of the index of stability of budgetary potential over 1.0 indicates a low level of budget stability, characterizes the persistence of current trends in the distribution of tax revenues and a decrease in the tax base for the main types of taxes, does not take into account the possibility of resettlement and economic development, it is characterized by a refusal to implement the mechanisms of sustainable and balanced spatial development of the Russian Federation. The main types of income are grants, subsidies, subventions.

Table 2 Rating of regions according to the value of the budgetary potential sustainability index, 2019 (fragment)

Regions	Total	Rank	Budgetary potential sustainability index	Level of stability of budgetary potential
Yamal-Nenets Autonomous Area	12.33	1	0.02	2
Moscow	13.52	3	0.06	2
Murmansk Region	14.54	8	0.16	2
Leningrad Region	14.60	9	0.18	2
St. Petersburg	14.62	10	0.20	2
Tula Region	15.90	28	0.57	1
Voronezh Region	15.91	29	0.59	1
Khabarovsk Territory	15.91	30	0.61	1
Nizhny Novgorod Region	**15.95**	**33**	**0.67**	**1**
Stavropol Territory	16.76	77	1.57	3
North Caucasian Federal District	16.78	78	1.59	3
Republic of Bashkortostan	16.80	79	1.61	3
Karachayevo-Circassian Republic	17.67	94	1.92	3
Republic of Altai	17.73	95	1.94	3
Republic of Tuva	17.82	96	1.96	3

Source Authors' calculations based on data from the Ministry of Finance of the Russian Federation [6] and the Federal State Statistics Service of the Russian Federation [2]

3 Results

The empirical results of calculations using the proposed method for determining the budgetary potential of the constituent entities of the Russian Federation and the level of stability of budgetary potential based on the optimality index are presented in Table 2.

4 Conclusion

Analysis of the results of approbation of the technique leads to the following conclusions.

- The first level of stability of budgetary potential (high stability) was assigned to regions that are increasing industrial potential, for example, the Nizhny Novgorod region became one of the leaders.

- The second level of stability of budgetary potential (sufficient stability) is possessed by the oil and gas regions.
- The third level of stability of the budgetary potential (low stability) is in the regions where there are no points of intensive industrial growth.

In general, the use of the developed methodological toolkit is important not only for the implementation of a targeted value-oriented state financial policy in the regions, but also for determining the budgetary strategy of the entire country, because it takes into account the individual characteristics of the development of each region from the standpoint of its resource industrial and economic potential and budgetary sustainability. The presented methodology allows, on the basis of the proposed levels of stability of budgetary potential, government bodies to improve tax and budget policies and develop a certain vector of their development in order to ensure the social well-being of citizens of the country.

Acknowledgements The study was carried out within the framework of the basic part of the state assignment of the Ministry of Education and Science of the Russian Federation, project 0729-2020-0056 «Modern methods and models for diagnosing, monitoring, preventing and overcoming crisis phenomena in the economy in the context of digitalization as a way to ensure the economic security of the Russian Federation».

References

1. Chigarin, A. Y. (2015). Ways to improve and increase the efficiency of the provision of educational services. *Problems of the Modern Economy, 3*, 357.
2. Federal State Statistics Service of the Russian Federation. (2019). https://rosstat.gov.ru/. Accessed 01 Dec 2021.
3. Fedorova, E. A., & Agadzhanyan, A. A. (2015). Regional spillover effect: An empirical analysis of the regions of the Russian Federation. *Audit and Financial Analysis, 2*, 283–287.
4. James, A. (2015). Is education really underfunded in resource-rich economies? Evidence from a panel of U.S. States. Working Papers. No. 2015–01. pp. 1–12.
5. Kornilov, D. A., Yashina, N. I., Yashin, S. N., Pronchatova-Rubtsova, N. N., & Vinnikova, I. S. (2019). Diagnosing changes in financial and economic indicators of the EU countries and the Russian Federation in crisis. *Journal of Advanced Research in Law and Economics, 9*(4), 1302–1311.
6. Ministry of Finance of the Russian Federation. (2019). https://minfin.gov.ru/ru/. Accessed 01 Dec 2021.
7. Porunov, A. N. (2017). Hermeneutics of the DEA-analysis methodology on the example of assessing the comparative effectiveness of the execution of the consolidated budget by the constituent entities of the Russian Federation in the field of preschool education. *Regional Economics: Theory and Practice., 15*(8), 1527–1539.
8. Saisana, M., Saltelli, A., & Tarantola, S. (2005). Uncertainty and sensitivity analysis techniques as tools for the quality assessment of composite indicators. *Journal of the Royal Statistical Society, 168*(2), 307–323.
9. Yashin, S., Yashina, N., Pronchatova-Rubtsova, N., & Kashina, O. (2018). Methodical approaches to assessing the budget potential of the region taking into account the innovative development of high-tech industries. In *European Financial Systems 2018. Proceedings of the 15th International Scientific Conference* (pp. 849–856). Brno: Masaryk University.

10. Yashina, N., Pronchatova-Rubtsova, N., Petrov, S., & Kashina, O. (2018). Methodical approaches to the formation of model budgets in order to improve the effectiveness of the budget process in Russia. In *European Financial Systems 2018. Proceedings of the 15th International Scientific Conference* (pp. 857–864). Brno: Masaryk University.

Big Data in the GovTech System: Future Perspectives and New Questions (Conclusion)

A systematic view of the GovTech, formed in this book, made it possible to substantiate the key role of Big Data in the provision of high-tech educational services in GovTech. The high efficiency of state regulation of various sectors of the Russian economy was also established with the help of Big Data in the GovTech. Several manifestations of the digital divide have been identified, aggravated by the COVID-19 pandemic and crisis, but successfully overcome in Russia with the help of GovTech based on Big Data.

This indicates the expediency of further development of the GovTech system based on Big Data, the prospects of which are associated with the digital modernization of the process of providing the entire range of public services; with the creation of "smart" cities, the state regulation of which is based on machine vision; with the completion of the transition to a digital government that operates primarily in an electronic environment. At the same time, along with new scientific knowledge obtained in this book, new questions have also been raised.

Among them is the ethical issue of the development of the GovTech system based on Big Data. While machine vision makes Big Data more accessible and empowering to the GovTech system, this innovation has clear societal implications that can be both positive and negative.

To mitigate the social consequences, a balance must be found between the desire of society as a whole to increase the effectiveness of the GovTech and the willingness of each individual to limit their non-state-controlled space (the boundaries of state intervention in private life). It is necessary to determine for each area and practice of management what is more important: a guarantee of compliance with the "rules of the game" or freedom of action. This will prevent social resistance to change during the development of the GovTech system based on Big Data.

The issue of financing the development of the GovTech system based on Big Data also deserves attention. The pandemic and the COVID-19 crisis have aggravated the shortage of financial resources in public budgets around the world. Therefore, each economic system will have to make a choice, which is more important for them: the

V. N. Ostrovskaya and A. V. Bogoviz (eds.), *Big Data in the GovTech System*, Studies in Big Data 110, https://doi.org/10.1007/978-3-031-04903-3

development of public digital infrastructure (public goods) or targeted (individual) social support (economic benefits).

Another issue is related to cybersecurity. The GovTech system based on Big Data should not only protect the personal data of citizens and organizations but also work smoothly. Since public administration has universal coverage in society and the economy, it is unacceptable for cybercriminals to hack the GovTech system based on Big Data or paralyze the operation of this system.

The updated issues also include the need for a balanced and sustainable development of the GovTech system based on Big Data. Territories remote (for example, rural or northern) from the capitals of the countries of the world should also be covered by the GovTech system, which, in particular, requires even more efforts for the digital development of these territories in Russia. It is also necessary to overcome the digital divide in all its manifestations. The future of GovTech lies in expanding the coverage of business practices with monitoring based on Big Data.

These issues require further more careful study—it is proposed to devote further research to them.

Printed in the United States
by Baker & Taylor Publisher Services